Creo Parametric Modeling with Augmented Reality

Creo Parametric Modeling with Augmented Reality

ULAN DAKEEV
Sam Houston State University, Texas, USA

Library of Congress Cataloging-in-Publication Data
Names: Dakeev, Ulan, author.
Title: Creo Parametric modeling with augmented reality / Ulan
  Dakeev.
Description: Hoboken, NJ : Wiley, 2023. | Includes index.
Identifiers: LCCN 2023012126 (print) | LCCN 2023012127 (ebook) | ISBN
  9781119894414 (paperback) | ISBN 9781119894445 (ebook) | ISBN
  9781119894421 (adobe pdf) | ISBN 9781119894438 (epub)
Subjects: LCSH: Creo Parametric. | Engineering models. | Three-dimensional
  imaging. | Augmented reality.
Classification: LCC TA177 .D35 2023 (print) | LCC TA177 (ebook) | DDC
  620.00285/53–dc23/eng/20230330
LC record available at https://lccn.loc.gov/2023012126
LC ebook record available at https://lccn.loc.gov/2023012127

Cover Design: Wiley
Cover Image: © Liyao Xie/Getty Images

Set in 10/12 STIXTwoText by Straive, Pondicherry, India
SKY10048136_051923

Table of Contents

CHAPTER 1

Introduction to Parametric Design

1.1 Introduction

Computer-aided design (CAD) is the development of new part models and modification of already existing parts' designs for new prototypes of complex assemblies such as cars, planes, computers, etc. The development and modification of parts include analysis of design data from customers and other stakeholders. Additionally, the CAD provides communication tools such as drawings and simulations to accurately estimate and manufacture customer's specifications.

This chapter introduces the basic concepts behind parametric design and feature-based modeling. Parametric design is a process used to manipulate, regenerate, and design objects based on a set of rules or parameters. There are many concepts that appear regularly in parametric design projects. One important concept is repetition, which works best when it is used in parallel with rotation, scaling, and movement. This simply means to copy an object multiple times while introducing a gradual change in its scale for example. Within this chapter, engineering graphics and three-dimensional modeling concepts will be explored. Additionally, parametric modeling principles will be covered. Upon finishing this chapter, you will be able to:

- Describe the utilization of computer-aided design within engineering graphics.
- Compare three-dimensional modeling techniques.
- Describe concepts associated with parametric modeling and design.
- Describe the use of parametric design within a concurrent manufacturing.

Creo Parametric Modeling with Augmented Reality, First Edition. Ulan Dakeev.
© 2023 John Wiley & Sons, Inc. Published 2023 by John Wiley & Sons, Inc.

CHAPTER OBJECTIVES

After completing this chapter, you should:

- Understand what Creo Parametric is.
- Understand associativity, design intent, feature based modeling, parametric modeling, and solid modeling concepts.
- Learn how design process occurs and how teams from various disciplines work together to accomplish their common goal
- Learn how to create a free account, download and install Creo Parametric software
- Understand various file types within Creo Parametric and how to use the Augmented Reality application that accompanies this book.

1.1.1   What is Creo Parametric?

Creo Parametric (formerly known as Pro/Engineer) is a suite of programs that are used in the design, analysis, and manufacture of a virtually unlimited range of products. In a nutshell, Creo Parametric is a parametric, feature-based solid modeling software.

1.1.2   Definitions

Associativity	The sharing of a component's database between its application modes.
Design intent	The intellectual arrangement of assemblies, parts, features, and dimensions to meet a design requirement.
Feature-based modeling	The building of a part model with intuitive and realistic features, such as holes, pads, pockets, fillets, and chamfers and leaving low-level geometry like lines, arcs, and circles for the Creo Parametric to figure out.
Parametric design	"Parametric" means that the physical shape of your part or assembly is driven by the attributes (dimensions) of its features. You can define or modify the attributes at any time and the model is regenerated throughout your model and assembly and drawings. Creo Parametric has **bidirectional associativity** – this means changing a dimension on a drawing changes the shape of the model and vice versa.
Solid modeling	The developed computer model contains all the necessary information that a real solid object would have.

This book discusses solid modeling of parts, developing technical drawings or blueprints for those parts with geometric dimensions and tolerances (GD&T), using these part components within a larger assembly file, developing assembly drawings with balloons and bill of materials, developing simulation reports for modeled parts with finite element analysis (FEA), and modeling sheet metal products. It is worth to recognize what file types Creo Parametric will generate for the type of required functions (Table 1.1):

TABLE 1.1   **Naming Conventions Used in Creo Parametric in this Book**

File Extension	File Type	Description
.sec	Section file	Saves sketches in .sec extension
.prt	Part file	Saves 3D modeled parts
.drw	Drawing file	Saves drawings of parts and assemblies
.asm	Assembly file	Saves assemblies of part components
.rwd	Simulate result definition file	Saves simulation results
.pbk	Mechanism playback file	Saves mechanism animation results
.mpeg	Moving picture experts group file	Saves captured video of mechanism animation

1.2   Introduction to Computer-Aided Design

Engineering design graphics has made significant changes since the early 1980s. For the most part, these changes are a result of the evolution of CAD. Before CAD, design was accomplished by traditional board drafting utilizing paper, pencil, and various other manual drafting devices.

1.2.1   Design Process

New design concepts or ideas should exist in the mind of the designer before they can be modeled. Working in a team continuous improvement process or a group of students in the classroom may generate a need to consult each other before starting the Creo Parametric. This will require clear expression of the ideas verbally, symbolically, and graphically. The design process is your ability to combine scientific principles, resources, and existing products into a problem solution. Suppose a customer called your company and said the tractor they purchased has stopped in the middle of a cornfield due to the broken part. A group of engineers will go to run the test and collect the data on the tractor. Once the data is collected, a team is formed with people who have specific skills. Each member will express their opinion regarding the problem. Set of quality control tools (such as why-why diagram or fishbone diagram) might be developed in order to conduct the root cause analysis. Notice that the identification of the problem and the brainstorming sessions are accompanied with evaluating the ideas and selecting the appropriate ones until the final possible solution is identified. Developing three-dimensional part and the drawing would be the next "implement" stage of the problem-solving phase. This stage needs to be checked and approved in order to manufacture a product. The design engineer is responsible for checking and approving the production drawings. Dimensions, tolerances, and manufacturing notes (such as material) must be present on the drawing and discussed with the supplier before sending for production.

1.2.2 Free Creo Download and Installation (Student Version)

Power to create (PTC) provides a free license for Creo Parametric for educational purposes. Although some of the functions are limited in the student version, all of the tutorials discussed in this book can be achieved with the free version. Because Creo Parametric is a Windows-based software, you can install Windows operating system on your Mac via Parallels software or through the Bootcamp.

Instructions to Download Creo Student Edition (Note: File size is approximately 4 GB for 64 bit)

1. **Use your web browser to navigate to:**

   https://www.ptc.com/en/academic-program/academic-products/free-software/creo-college-download

2. **Enter the following information to start the process to acquire the software:**

   First (given) name: (*your name*)

   Last (family) name: (*your name*)

   Email address: (*your SHSU email address*)

   City: Where your school is located, i.e. Huntsville

   State: Where your school city is located, i.e. Texas

   Country: Where your school city is located, i.e. United States

   School/Institution: Name of your school, i.e. Sam Houston State University

   Grade level: Your school level, i.e. Undergraduate

   How will this software be used? In Classroom

   How many computers will you install the software on? 1

   How did you find out about this software? Professor

3. **Installation instructions**

   Follow the link to download CREO version your professor recommended. This link will lead you to a PDF with the instructions on how to proceed. *This PDF is important since it is your PRODUCT CODE, keep the PDF open (or save it if you need to repair or reinstall your software).*

   Follow the link on the PDF to download PTC CREO version, which also lets you CREATE A NEW ACCOUNT. To create an ACADEMIC ACCOUNT, enter the following information:

   First name: (*your name*)

   Last name: (*your name*)

   Email: (*your SHSU email address*)

   Academic program: Other

   Title: College/University Student

   School: Your School name, i.e. Sam Houston State University

   State/Province: Your Province or State, i.e. TX

   Country: Your country of residency, i.e. United States

Create a password. DON'T FORGET IT. Then enter the CAPCHA into the box that are underlined and to the right of the *PTC Security Challenge* box.

4. With your account created, go back to the *Quick Installation Guide* PDF that PTC created for you and follow the instructions to download CREO version.

1.2.3 Creo Installation

- Once the download complete, extract (unzip) the download folder ➔ Click "Setup" (Figure 1.1)

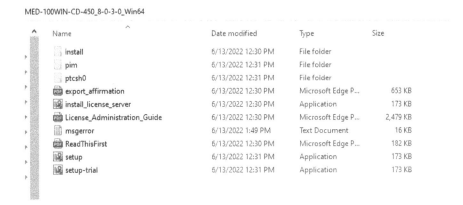

FIGURE 1.1   Creo Installation Folder

- Select "Install new software" (unless you are adding packages) ➔ Click "Next" on the Creo Installation Assistant Window (Figure 1.2)

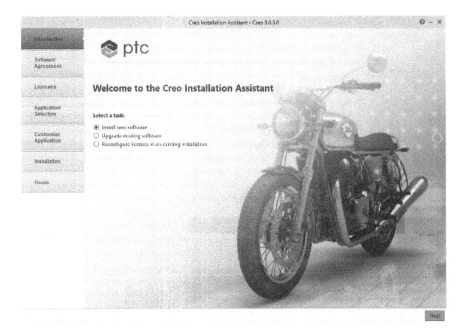

FIGURE 1.2   Creo Installation Assistant

- Read through the Software License Agreement ➔ Accept the license agreement ➔ Click Next (Figure 1.3)

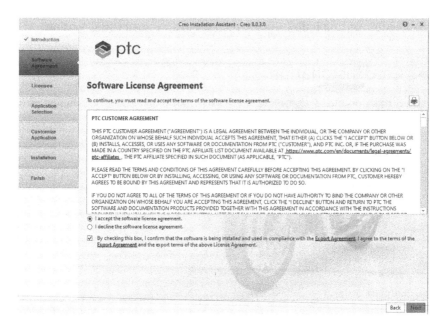

FIGURE 1.3 Creo License Agreement

- On the License Identification window (Figure 1.4), enter the serial number on your PDF document (Quick Installation Guide) you opened/saved earlier (the serial number will look like BK600908EDSTUDENTEDUNI)

FIGURE 1.4 Creo Installation Assistant

- Enter your PTC username (the one you created earlier) and password → Click Login
- Once you see "Success," click Next to start Creo Parametric and packages

1.2.4 Starting Creo Parametric

Previously known as ProEngineer Wildfire, Creo Parametric received its name in 2011 to include additional applications. The user interface of Creo Parametric is set up similar to all windows applications that comprises of the ribbon with most frequently used tools, menu tabs for additional tools, and the workspace, where all the modeling and sketching takes place. Initial start of Creo Parametric launches a welcome screen (Figure 1.5) with basic functions.

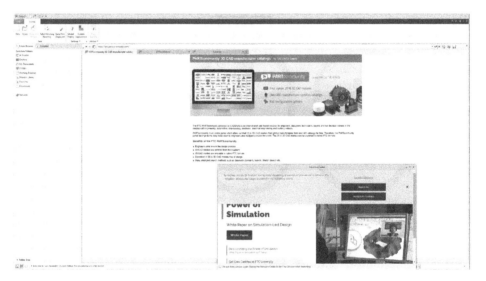

FIGURE 1.5 Creo Parametric Welcome Screen

Because Creo Parametric uses context-sensitive application menus, available menus can be used right away, while grayed out menus or options will become available as applicable. For instance, the "Open Last Session" tool is grayed out and is not available on Figure 1.5. This means we just started Creo Parametric and do not have an active last session. On the other hand, the "New, Open, Select Working Directory, etc." tools are available for use. Later in Chapter 2, we will start our first model with clicking the "New" button and myriad of additional tools will be available for use. The left browser panel lets the user navigate through local files and folders, additionally, expanding the "Folder Tree" provides more navigation options. Once the part modeling starts, this left panel will become a "Model Tree," where each modeling step is presented. Initially, the welcome screen contains built-in internet browser that you can even watch on YouTube videos. The purpose of this browser is, of course, not for YouTube videos but navigate through the database called PDMlink within the organization to check out and check in parts. For instance, you as a new design engineering in a manufacturing company revised one of the parts for future model car and save it. But where did you save? On your computer? How will other people have access to the part when you are away? Yes, you could send them with an email, but that could take up email storage space. Therefore, the PDMlink database lets organizations check in (upload) their parts onto the server, which will instantly become available for other team members regardless their location in the world. This enables organizations to stay operational 24/7 as people from China or India can pick up your last modified work and continue as you go home after your shift in the States. Similarly, you will pick up the updated iteration of the model next morning with significant amount of progress on the part.

Although it is possible to use laptop's touchpads to model some basic 3D models in Creo Parametric, it is highly recommended and crucial to use a three-button mouse (left mouse button, right mouse button, and the scroll wheel, as referred to middle mouse button) as the middle mouse button can end a command or use for panning and zooming functions. Below (Figure 1.6) is the explanation of what mouse button functions are in Creo Parametric:

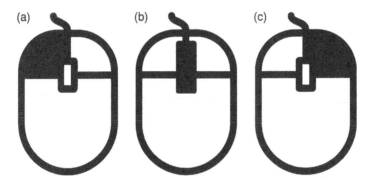

FIGURE 1.6   (a) Left Mouse Button (LMB), (b) Middle Mouse Button (RMB), (c) Right Mouse Button (RMB)

LMB (left mouse button): Mainly used to select tools, commands, and operations.

MMB (middle mouse button): A very handy tool to accept or end currently selected tool/operation or zooming in/out and rotating/spinning/panning the geometry.

RMB (right mouse button): Majority of times the right mouse button is used with clicking and holding the button, which brings quick access to most frequently used tools and commands.

It is a good practice to organize the folder before starting work with Creo Parametric. It is highly recommended to set up a working directory, so the saved files are stored in one place. Although setting up a working directory is discussed in Chapter 2, it is good to understand why we need to set it up. Consider a larger assembly file that comprises of 5 components. This assembly could be your pen and the components are as follows: 1, lead; 2, pen housing; 3, cap; 4, spring; and 5, button to retrieve the lead. Let us assume that all these components are saved in different locations of the computer, such as my documents, miscellaneous on drive d, or on a usb thumb stick. You can start a new assembly and locate all of these components from their stored locations and your assembly will be complete. However, the moment you send the assembly to someone else or you want to open it on a different computer, the assembly will fail to open its components. This is because the folders that contain the components are inaccessible by the new computer, and it does not know where they are. Therefore, saving all those components in a single working directory will automatically retrieve all assembly components on a different computer. If you are using learning management system such as Blackboard, uploading a folder with all parts, drawings, and assembly files needs the folder compressed. To compress a folder, Right click→Send to→Zipped/compressed file (windows), on Mac, right click and compress.

1.2.5 Augmented Reality Companion Application and Image Targets

This book is accompanied with an Augmented Reality (AR) application that you can download and install on your Android (in development for iOS) devices. The application enables to visualize and understand the three-dimensional representation of the blueprint or the outcome of the part, assembly, or animations. The AR application provides interactive buttons to rotate, scale, assemble, and watch a short, animated clip of a tool. Simply start the application, point your Android device's camera on the figures within the textbook to experience the AR.

For effective use of the AR application, install the application on your Android device through the following QR code (Figure 1.7), open the application and point your device's camera to the image target.

FIGURE 1.7 QR Link to Download and Install the AR Companion App

Each chapter has an opening figure that can be augmented with the application (Figure 1.8). All image targets are compiled separately in the Appendix for reference. You can rotate each 3D model from the AR application to visualize and inspect various features of parts from different orientations. This version of

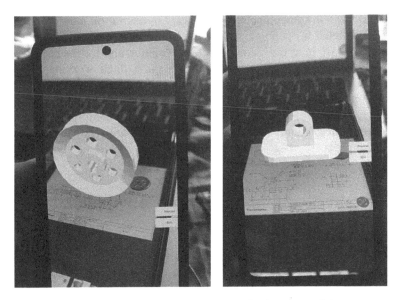

FIGURE 1.8 AR Companion App Projecting 3D Models

the AR companion application augments the completed part before starting the chapter. Animated tutorial clips are excluded from this version to reduce stress on the mobile device. Either all animated tutorials will be delivered as an update to the companion application or the user will be able to install it separately through the same QR code.

CHAPTER 2

Introduction to Sketch

2.1 Introduction

> **[AR]** Symbol indicate you can use your mobile devices to interact augmented reality.

This chapter introduces the basic sketch drawings. Users can quickly sketch their ideas.

CHAPTER OBJECTIVES

- Introduction to 2D sketcher
- Sketch modifications
- Sketcher constrains
- Sketch accuracy
- 3D model from 2D sketch
- Sketcher exercise

> **Sketch** is 2D drawing created on specific planar reference to develop 3D objects. In sketch mode, create various section geometry (line, rectangle, circle, arc, etc.), add dimensions and constraints.

For file management purposes, a new part will be developed and saved, opened, and closed.

Creo Parametric Modeling with Augmented Reality, First Edition. Ulan Dakeev.
© 2023 John Wiley & Sons, Inc. Published 2023 by John Wiley & Sons, Inc.

E1 **EXERCISE 1** | L-Shaped Flat Bracket

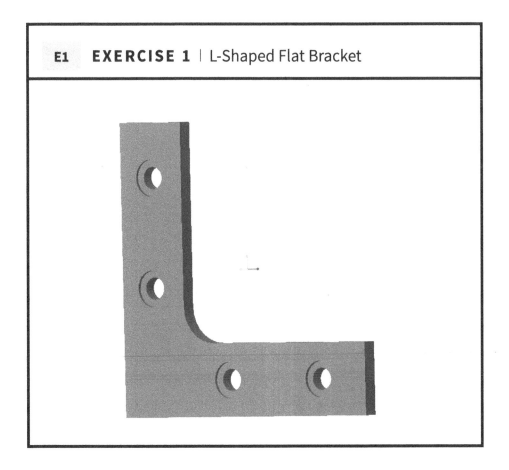

2.1.1 Select Working Directory

Working Directory

➢ Keeping files (parts, assemblies, drawings, etc.) organized increases efficiency in modeling.

Although Creo Parametric can retrieve necessary files from different locations, you are capable of setting a designated location to keep files in. The working directory is usually the modeling point for all Creo Parametric objects. When a new file is saved, it is saved in the current working directory, unless a new directory is specified.

- Select **"Select Working Directory"** (Figure 2.1) → Select Desktop (or other destination folder) → Create New Folder (by right click ⟳ on desktop area) → Name "Creo6_WorkingDirectory" → Click OK.

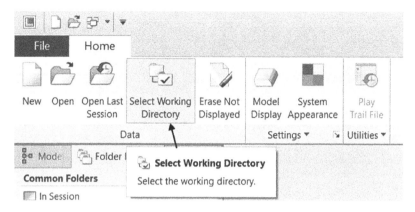

FIGURE 2.1 Working Directory

You have just set up a working directory named "Creo6_WorkingDirectory" on the desktop. You will notice that message area displays that the working directory is successfully set up.

Files will be saved in the selected working directory. You can always change the working directory destination.

2.1.2 Starting New Project

Select **"New"** to start modeling (Figure 2.2).

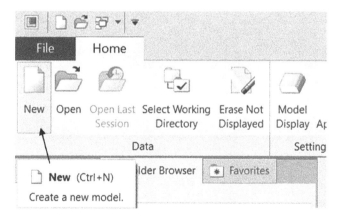

FIGURE 2.2 Click "New" to start a new project

You are presented with new dialog box with number of options to choose from Figure 2.3. Ensure "Part" is selected.

- Enter the part name (you may call it "L-Shaped_Flat_Bracket").
  Note: Do not leave space between characters.
- Uncheck "Use default template" → Select OK

FIGURE 2.3 Model Selection

From here, a new dialog box "New File Options" (Figure 2.4) will appear. Here you may specify units for the part to be modeled. In this exercise, we will choose inlbs_part_solid as template.

- Select "inlbs_part_solid" → Select OK.

FIGURE 2.4 Unit Selection

Work environment presented that the interface slightly changed and there are new buttons on the ribbon (Figure 2.5). Prior to creating any feature, you need to select a designated datum plane. By selecting the planes on the datum equally highlights itself on the model tree. Besides the ribbon on top of the Creo window, floating toolbar located on top of the workspace brings more options for displaying the model with shortcuts, such as zoom in/zoom out, rendering option, simulate, and so on.

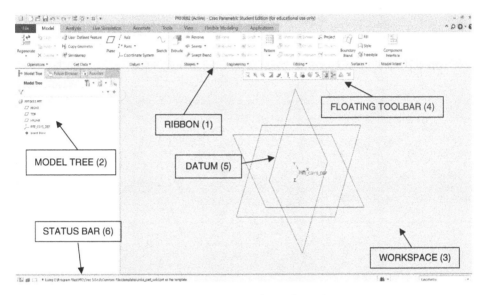

FIGURE 2.5   Creo Parametric Work Environment

Ribbon (1)	The ribbon is a user interface element located on top of the workspace (Figure 2.5) and contains most frequently used tools.
Model Tree (2)	List of features and sketches occurring in the part in current session.
Workspace (3)	The main working area displays the part and allows users to edit part during session.
Floating Toolbar (4)	The floating toolbar located on middle top of workspace with shortcuts, including zoom-in/zoom-out, display options, rendering options and simulate, etc.
Datum (5)	There are three datum planes by default.
Status bar (6)	The lower data panel containing message pane lists informational, debug, warning, and error messages logged during your session.

2.1.3 Start New Sketch

- Click "Sketch" tool from the ribbon (Figure 2.6) → Select "Front" plane → Click "Sketch" button on the dialog box.

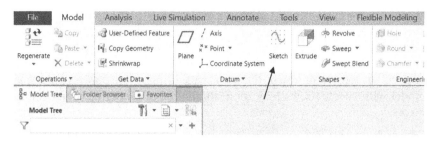

FIGURE 2.6 Sketch Tool

On the Display Toolbar click "Sketch View" 🖥 *button (Figure 2.7) in order to activate the sketching plane to be oriented parallel to the screen.*

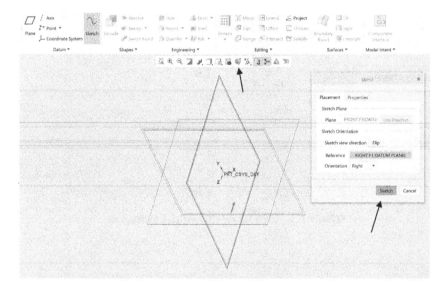

FIGURE 2.7 Select a Datum to Initiate Sketch

Select "Line" tool and start Sketch ⌐ shape as shown in Figure 2.8.

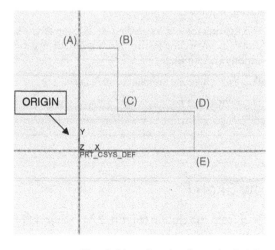

FIGURE 2.8 Sketch Lines Starting from the Origin

- Click on the Origin, continue to draw the line by clicking points (A, B, C, D, E) in sequence, and back to the origin (*you do not need to keep left mouse button pressed while moving*).
- To complete the sketch, press middle mouse button 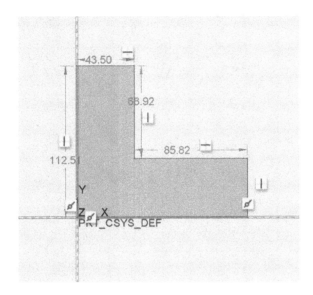 or press ESC twice until the sketch is shaded as in Figure 2.9.

FIGURE 2.9   Creo Automatically Adds Weak (Driven) Dimensions on Objects

---

Before leaving a sketch, you must make sure the sketch is **fully defined** – which includes:

✓ The sketch is shaded,
✓ All dimensions are strong dimensions.

Weak dimension – Strong dimension
The dimension are determined by users and indicated by two colors (Figure 2.9):

➢ Weak dimension is in aqua/light blue color
➢ Strong dimension is in purple (dark blue in Creo 5.0/ProEngineer and earlier versions).

2.1.4 Add Constraints

Constraints are conditions applied to define the geometry or relationship among entities while drawing sketch. There are eight constraints introduced in Creo: **vertical, horizontal, perpendicular, tangent, midpoint, coincident, symmetric, equal,** and **parallel**. Some constraints can be applied to a single entity, while other constraints can be applied to groups of more than two entities.

Applying constraints can help to efficiently ease the process of defining a sketch.

- Select the "Equal" command on the **Constrain** tab (Figure 2.10). Then select two pairs of lines – a with b and c with d, respectively.

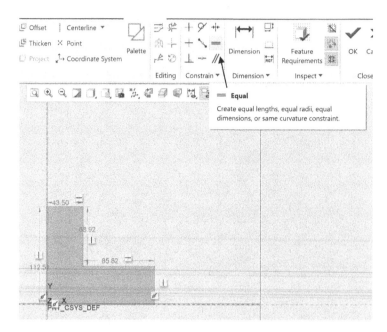

FIGURE 2.10   Use Equal Constraint When Lines Are Equal Length

Note: As you apply the constraints, the system will automatically eliminate weak dimensions.

2.1.5   Define and Modify Dimensions

- Click "Dimension" tool from the Ribbon (Figure 2.11) → Select the bottom horizontal line → Move the cursor down to the blank space underneath → Press middle mouse button ⊖ to define the strong dimension shown in purple color (Figure 2.12).

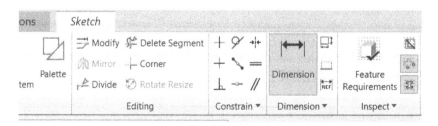

FIGURE 2.11   Dimension Tool

- Make a rectangular selection by pressing left mouse button and dragging the mouse cursor to include the whole sketch and dimensions (Figure 2.12).

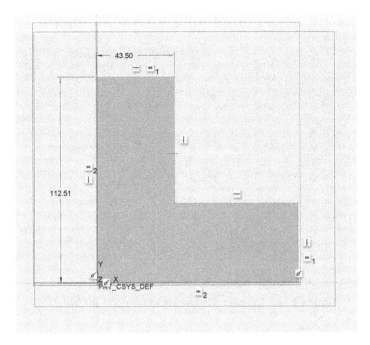

FIGURE 2.12   Strong Dimensions Are Defined

- Select "Modify" tool from the Ribbon (Figure 2.13)

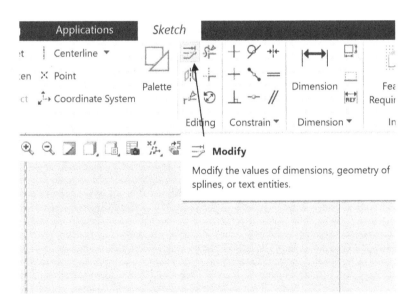

FIGURE 2.13   Use Modify Tool to Enter Values at Once

- A dialog box (Figure 2.14) will appear shortly with adjustable dimensions → Enter the value as shown in Figure 2.14. Click OK to generate the shape following the dimensions.

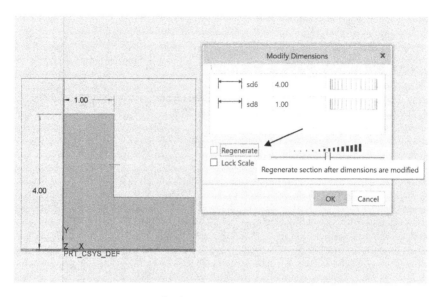

FIGURE 2.14 Uncheck Regenerate to Keep Shape Unaltered

Note: Uncheck "Regenerate" (checking "Regenerate" will update shape in real time).

2.1.6 Create Holes with Sketch

Select "Circle" tool in Sketching tab to create circle features on the sketch.

- Draw a circle inside the sketch as shown in Figure 2.15 → Continue to draw another circle underneath.

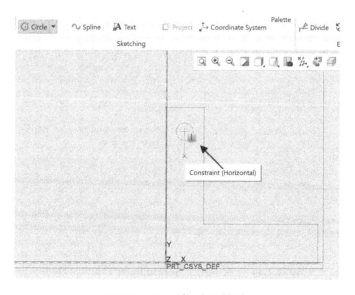

FIGURE 2.15 Alinging Circle

Note: Make sure circles are horizontal and vertical aligned, as you move the cursor straight downward, the horizontal constraints are automatically applied (Figure 2.15).

- Repeat drawing four circles (Figure 2.16).

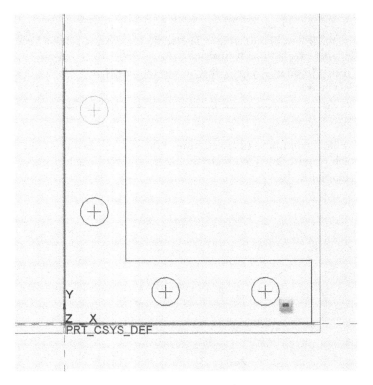

FIGURE 2.16 Constraints

Note: Equal constraint will be automatically indicated as shown in 2.18 when drawing circles.

- With dimension tools, define strong dimensions and modify them as shown in Figure 2.17. Click OK to generate the shape with defined dimensions.

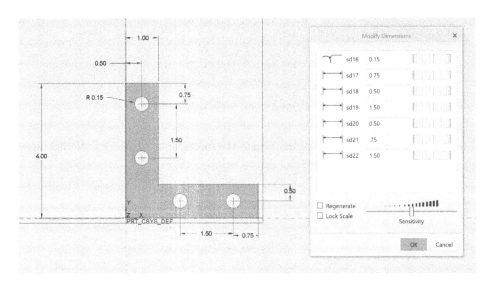

FIGURE 2.17 Strong and Accurate Values

2.1.7 Fillet Command

> **Fillet** is a round feature at the corner or intersection of two non-parallel lines. Creating a fillet will connect two entities with an arc.
>
> Besides Fillet, **Chamfer** is also a tool to create a beveled corner connecting two non-parallel entities.

- Select "Fillet" command in "Sketching" tab → Select two lines at the corner of the sketch (Figure 2.18) to create the fillet → Click Select or press ESC to exit sketching mode.

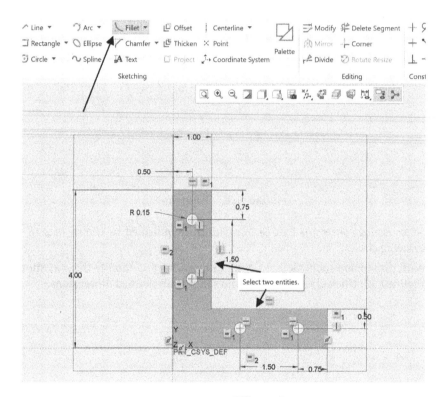

FIGURE 2.18 Fillet Tool

- Double click to the radius dimension of the fillet (Figure 2.19) → Enter **0.5** in. → Press to result Figure 2.20.

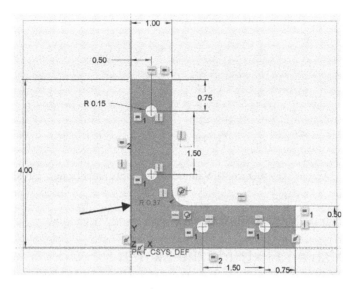

FIGURE 2.19 Define Weak Dimension

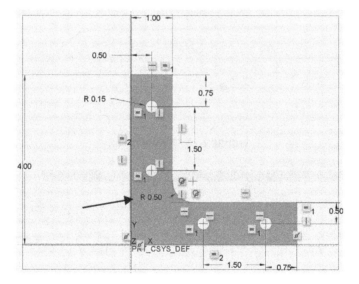

FIGURE 2.20 Strong Dimension and Value

- Now, as the sketch is fully defined and shaded, you can click on the green check mark on top right corner to accept sketch and exit sketch mode (Figure 2.21).

FIGURE 2.21   Accept Sketch Once All Values Are Accurate and Strong, with Shaded Shape

*Note: Verify the sketch **fully defined** before leaving sketch mode.*

2.1.8   Finish and Extrude the Sketch

- Select "Extrude" tool on the ribbon in "Shapes" tab (Figure 2.22). New tab will appear as Figure 2.23.

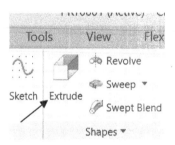

FIGURE 2.22   Extrude Tool

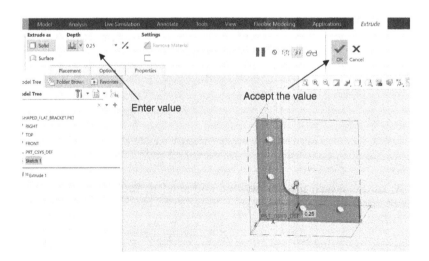

FIGURE 2.23   Enter Extrude Value

Note: If "Extrude" command is not working, recheck and verify "Sketch 1" on the model tree is active.

- Enter **0.25** in. extrusion value (Figure 2.23) → Click on the green check mark on the top corner to accept the geometry. The part will be completely extruded as shown in Figure 2.24.

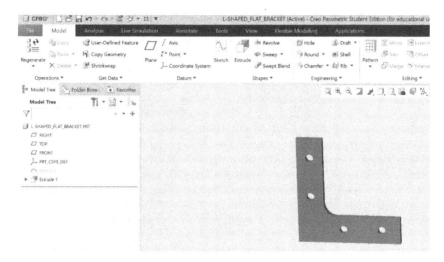

FIGURE 2.24   Review Complete Part

- Select Save on top left corner or press Ctrl+S to save the part → Close the window by selecting ⊠ on the top left.

Note: Selecting ⊠ on the top right will result in erasing current sessions and terminating Creo software completely.

2.1.9   Sketch Counterbore

- Make additional 0.25-in. radius circles as shown below (Figure 2.25).

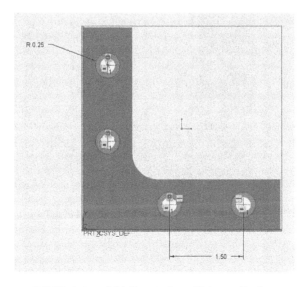

FIGURE 2.25   Add Counterbore Holes on Surface

- Extrude 0.05 in. and select remove material (Figure 2.26). To change extrude direction, click arrow to remove material from the part.

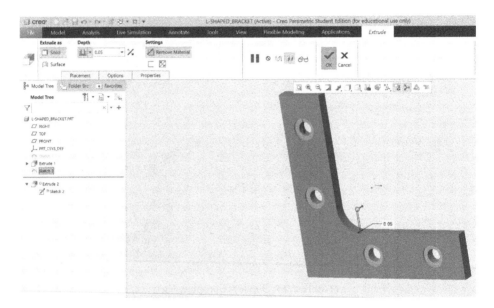

FIGURE 2.26 Extrude to Remove Material

- Select "ok" to complete exercise 1 (Figure 2.27).

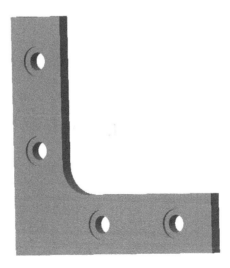

FIGURE 2.27 Review Completed Part

Chapter Problems

P2.1 Develop U-shaped part (Figure 2.28)

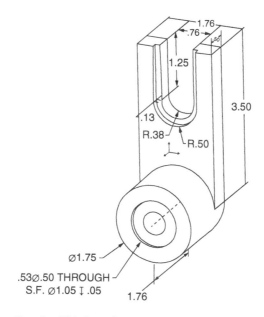

FIGURE 2.28   Develop This Part If Your Student ID Ends with ODD Number

P2.2 Model hinge bracket (Figure 2.29)

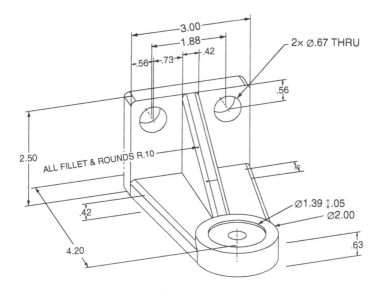

FIGURE 2.29   Build This Part If Your Student ID Ends with EVEN Number

P2.3 Base (Figure 2.30)

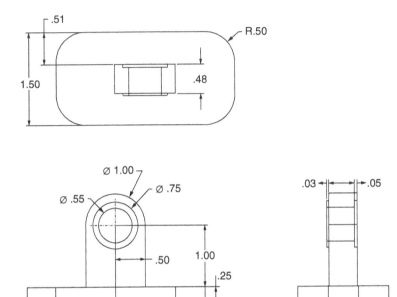

FIGURE 2.30 Projected Orthogonal Views for Base

P2.4 Link (Figure 2.31)

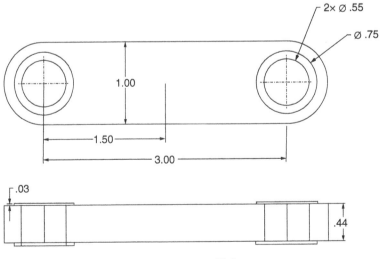

FIGURE 2.31 Link

P2.5 Pin (Figure 2.32)

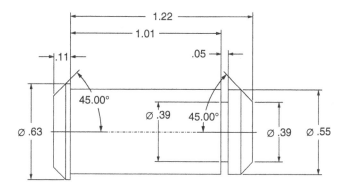

FIGURE 2.32 Pin

P2.6 Retaining ring *(Thickness: 0.5 in.)* (Figure 2.33)

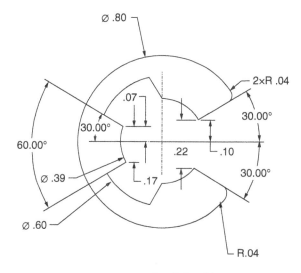

FIGURE 2.33 Retaining Ring

CHAPTER 3

Revolve and Sweep

3.1 Introduction

Definitions

A Revolved feature has a shape around a straight axis that cannot be bent (Figures 3.2–3.4).
Sweep, on the other hand, has a trajectory line (axis) that can be controlled, and a cross-section shape will follow this trajectory line (Figure 3.1).

There are numerous revolved objects such as pipes, cylinder objects, pencils, bowling pins, bottle

CHAPTER OBJECTIVES

After completing this chapter, you should:

- Why Revolve tool is used.
- Common objects that were designed with revolves.
- Difference between "Construction" and "Geometry" Center lines.
- Sweep Tools to control Centerline.
- Constraint options for sweep cap endings.

Creo Parametric Modeling with Augmented Reality, First Edition. Ulan Dakeev.
© 2023 John Wiley & Sons, Inc. Published 2023 by John Wiley & Sons, Inc.

E1 **EXERCISE 1** | Wind Turbine Shroud (To Improve Power Generation)

Revolve features are created in a similar way to Extrude features (*Chapter 2*). The main differences are that an axis of revolution (centerline) is required, and an angle of revolution is defined instead of depth in Extrude.

- Select new working directory and start new part (Chapter 2: p. 2).
- Name the part "Wind_Turbine_Shroud."

  Note: We will choose inlbs_part_solid *as template.*

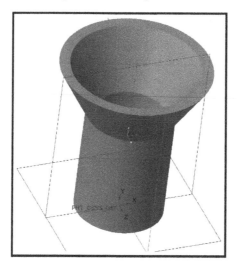

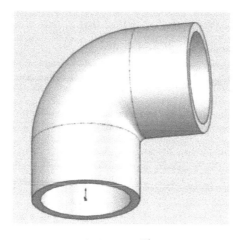

FIGURE 3.1 Pipes

FIGURE 3.2 Bowling Pins. *Source:* Envato Elements Pty Ltd.

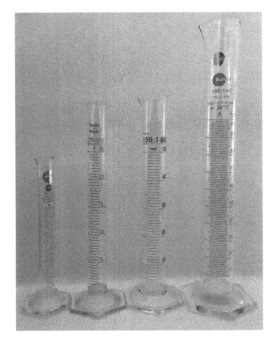

FIGURE 3.3 Graduated Cylinders. *Source:* Praphai Donphaimueang / Wikipedia Commons / CC BY-SA 4.0.

FIGURE 3.4 Cans

3.1.1 Start New Sketch

- Click "Sketch" tool from the ribbon (Figure 3.5) → Select "Front" plane → Click "Sketch" button on the dialog box.

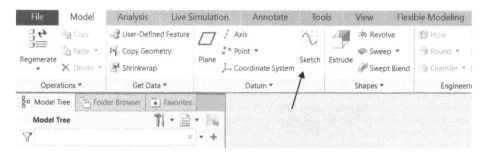

FIGURE 3.5 Sketch Tool

- On the Display Toolbar, click "Sketch View" button (Figure 3.6), activate the sketching plane (front) to be oriented parallel to the screen.

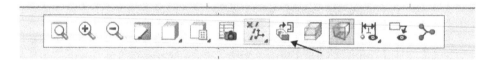

FIGURE 3.6 Sketch Orientation Parallel to screen

3.1.2 Select Centerline

You will notice two centerlines on the ribbon, we are interested in a geometry centerline (Figure 3.7) as this will aid with the 3D revolved feature, while the construction centerline can be used for references within the sketch.

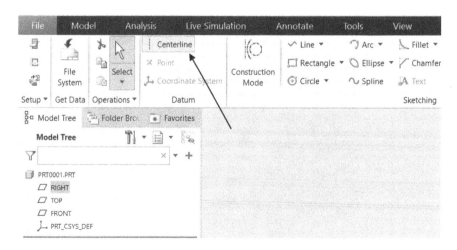

FIGURE 3.7 Geometry Centerline

- Click "Centerline" tool from the ribbon (Figure 3.7) → draw a straight dashed orange vertical line (Figure 3.8) → Hit Esc.

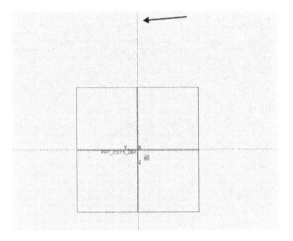

FIGURE 3.8 Geometry Centerline in Sketch

Select "Line" tool and start Sketch as shown in Figure 3.9.

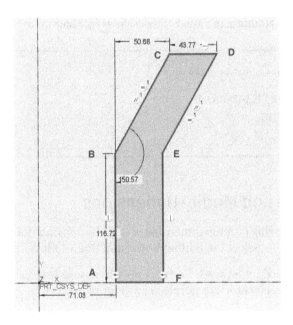

FIGURE 3.9 Sketch Sequence

- Click on the (A), continue to draw the line by clicking points (B, C, D, E, F) in sequence, and back to the origin *(you do not need to keep left mouse button pressed while moving)*.
- To complete the sketch, press middle mouse button ⟳ or hit ESC twice until the sketch is shaded (Figure 3.10).
- The reason why sketching far away from the centerline is there need to be a hole in the middle.

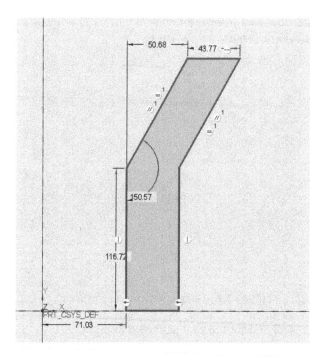

FIGURE 3.10 Weak Dimensions (Light Blue Color)

The following constraints should be applied:

- Parallel: AB and EF, BC and DE
- Equal: BC and DE
- Horizontal: AB, EF

3.1.3 Define and Modify Dimensions

- To define a driving (strong) dimension, Click "Dimension" tool from the Ribbon (Figure 3.11) → Select the bottom horizontal line → Move the cursor down to the blank space underneath → Press middle mouse button 🖱 to activate the strong dimension shown in purple color (Figure 3.12).

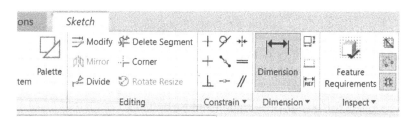

FIGURE 3.11 Dimension Tool

Note: We will need to ensure the shape is "shaded" and all dimensions are "strong" before we leave sketch. So far, we have ensured the sketch is shaded, and the

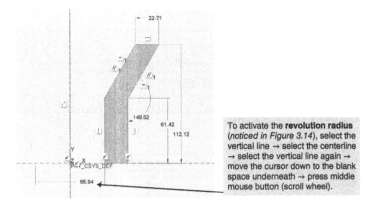

To activate the **revolution radius** *(noticed in Figure 3.14)*, select the vertical line → select the centerline → select the vertical line again → move the cursor down to the blank space underneath → press middle mouse button (scroll wheel).

FIGURE 3.12   Strong Dimension (Purple Color)

dimensions are strong. However, we need to enter the accurate dimension values per customer specification. To modify dimension values:

- Make rectangular selection (ensure select tool is active) by pressing left mouse button and dragging the mouse cursor to include the whole sketch and dimensions.
- Click "Modify" tool on the Ribbon (Figure 3.13) to open the dialog box (Figure 3.14).

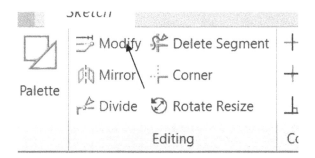

FIGURE 3.13   Select "Modify" tool to edit

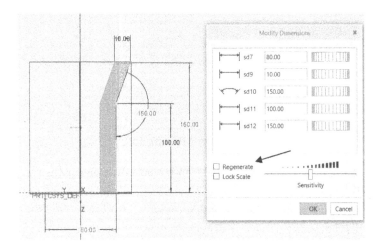

FIGURE 3.14   The Sketch Must Be Shaded and Driving Dimensions Should Be Accurate

On the "Modify Dimensions" dialog box (Figure 3.14):

- Uncheck "Regenerate"
  Note: Check Regenerate will update shape in real time.
- Enter accurate dimension values (Figure 3.14)

Click on the green check mark on top right corner to accept sketch (Figure 3.15).

FIGURE 3.15   Accept Sketch

3.1.4   Revolve

- Select **Revolve** command (Figure 3.16).
  Note: Make sure that Sketch 1 on Model Tree is selected.

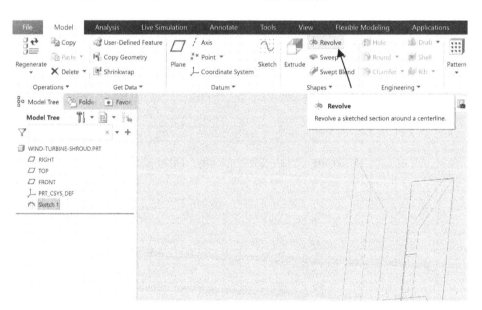

FIGURE 3.16   Revolve Tool

- Enter the angle of revolution as **360.0** and select **OK** to accept the revolve feature (Figure 3.17).
  Note: You may spin/rotate the shape (hold scroll wheel and drag) to see 3D revolved part.

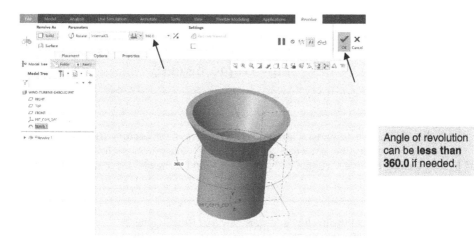

FIGURE 3.17   Revolved Feature to 360°

Angle of revolution can be **less than 360.0** if needed.

The part will be completely extruded as shown in Figure 3.18.

- Select Save ⊞ on left top corner or press Ctrl+S to save the part. Close the window by selecting ⊠ on the top left.

Note: Selecting ⊠ *on the top right will result in erasing current sessions and terminating Creo software completely.*

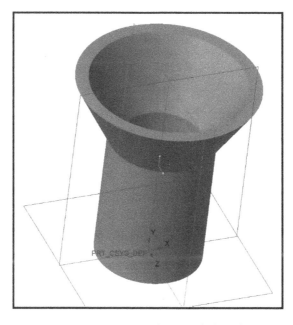

FIGURE 3.18   Complete Wind Shroud

E2   **EXERCISE 2** | Modeling a Radiator Coolant Hose
Pipe (Topic: Sweep)

Sweep features are created with an axis of references (**trajectory line**) and a defined sketch as the cross-section geometry. Unlike revolve where center-line is straight, sweep "centerline" (trajectory line) can be bended and adjusted as desired.

- Select new working directory and start new part (see Chapter 2).
- Name the part "Sweep-pipe."

  Note: We will choose inlbs_part_solid *as template.*

3.1.5   Start New Sketch

- Click "Sketch" tool from the ribbon → Select "Front" plane → Click "Sketch" button on the dialog box.
- Select "Circle" tool (Figure 3.19) and Draw 2 circles with equal diameter/radius (Figure 3.20).

  Note: Equal constraint will be automatically applied when the diameters/radiuses are drawn to be equal. If not, you should set equal constraints after drawing the sketch.

FIGURE 3.19   Radiator Hose

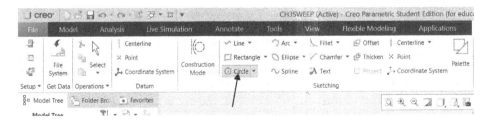

FIGURE 3.20   Circle Tool

- Select "Line" tool in "Sketching" tab → Draw a line randomly connect these two circles (Figure 3.21).

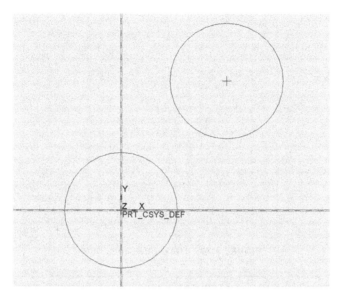

FIGURE 3.21   Sketch Two Equal Diameter Circles

- Select "Tangent" tool in "Constrain" tab → Select the line and each circle once at a time to apply tangent constraint for the line and 2 circles (Figure 3.22) → Hit ESC twice to return to Select mode.

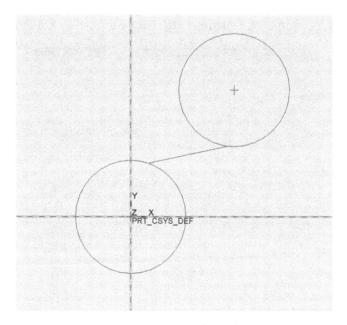

FIGURE 3.22   Connect Two Circles with Tangent Line

- Select "Horizontal" constraint tool and select two endpoints of the line. This will make the two ends aligned horizontally, also brings two circle edges to the same level (Figure 3.23).

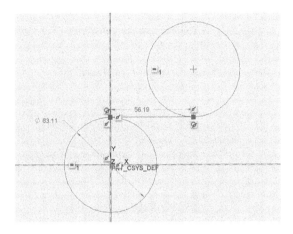

FIGURE 3.23 Horizontal Constraint

The following constraints should be applied:

- Equal: 2 circles' diameter/radius.
- Tangent: line to 2 circles.

Note: See Chapter 2 to apply constraints.

- Select construction "Centerline" tool in Sketching tab (Figure 3.24) → Create 4 centerlines inside 2 circles, parallel to ZY and ZX axis as shown in Figure 3.24.

Note: These construction centerlines will disappear when we exit sketch mode.

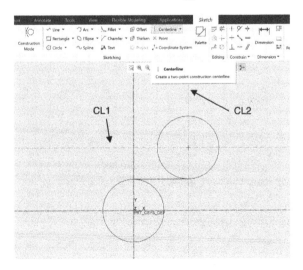

FIGURE 3.24 Trim Excessive Lines (Use Delete Segment tool)

- Select "Delete Segment" (Figure 3.25a) to trim excessive arcs and lines.
- Delete excessive lines by clicking the lines to get the final sketch as Figure 3.25b.

(a)

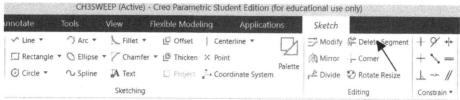

(b)

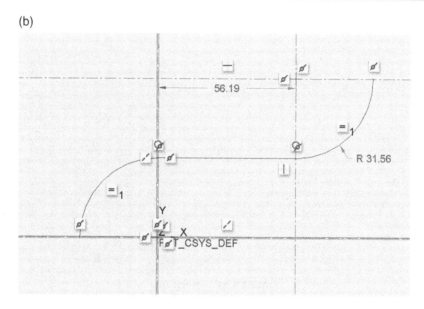

FIGURE 3.25   (a) Delete Segment Tool, (b) Trajectory Line with Weak Dimensions

- First, define driving (strong) dimensions, then modify them to correct values as shown in Figure 3.26.
- Verify all strong dimensions are present for the trajectory line and exit the sketch mode.

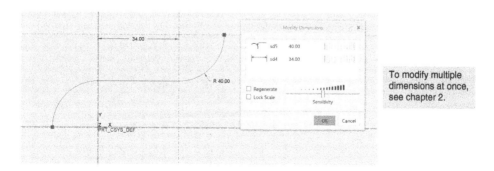

FIGURE 3.26   Trajectory Line with Strong Dimensions

To modify multiple dimensions at once, see Chapter 2.

Now that we have completed sketching the trajectory line, we need to sketch a cross-sectional geometry that follows our trajectory line. To sketch a cross-sectional geometry:

- Select "Sweep" (Figure 3.27).

  Note: Verify Sketch 1 (trajectory line) is selected on the Model Tree.

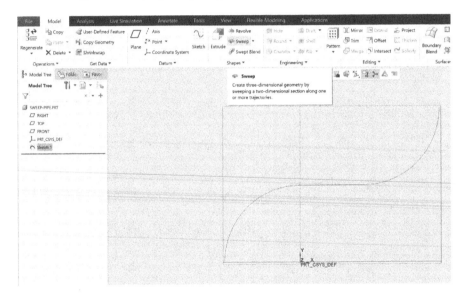

FIGURE 3.27 Sweep Tool

- Click Sketch tool (Figure 3.28) to define the cross-section geometry.

  Note: You will notice an arrow at one of the endpoints of the trajectory line. This indicates your sketch position of the cross section. If you want to select the other end, click on the arrow to see it change its location.

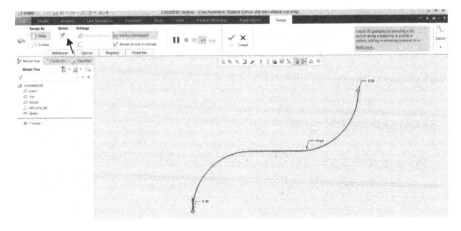

FIGURE 3.28 Select Trajectory Line

- Select Sketch View command to orient the sketch plane parallel to user interface (Figure 3.29).
- Select "Circle" tool → Create 2 circles, with 12" and 10" diameter, as shown in Figure 3.29.

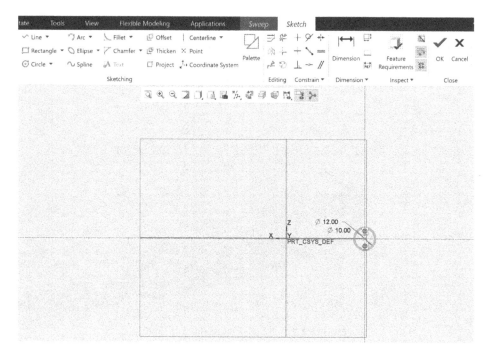

FIGURE 3.29   Sketch Two Circles for Inner and Outer walls

- Click OK to accept the cross-section sketch (Figure 3.30) and get back to Sweep mode.

  Note: You may need to rotate to observe the pipe from various orientations

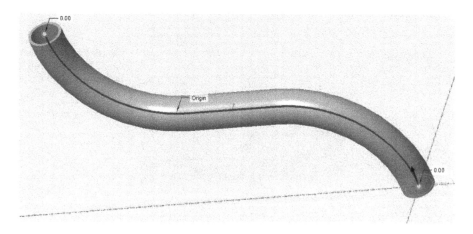

FIGURE 3.30   Cross-sectional Geometry with the Trajectory Line

- Click OK to accept the feature and exit Sweep mode.

3.1.6 Create the Flap Flanges

Although the radiator hose development is complete with the last step (Figure 3.30), we would like to discuss how we can add flanges to both ends for fastening purposes to flat surfaces. Similar geometries can be found at swimming pool rail handles. To add flanges to the pipe:

- Select "Sketch" tool → Select the surface at the end of the pipe (Figure 3.31).

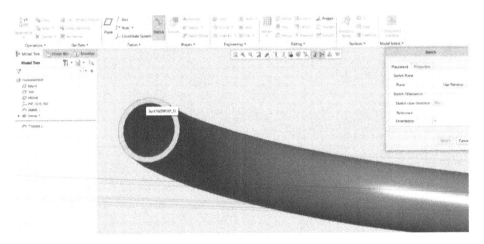

FIGURE 3.31 Select Flat Surface of Pipe

- Select "Circle" tool → Sketch 2 circles as shown on Figure 3.32.

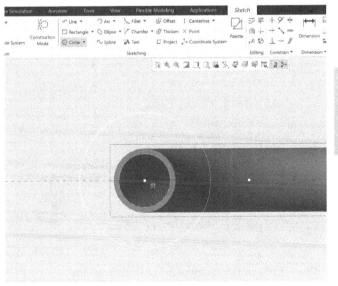

The centers of 2 circles should be **coincident** to the end point of the pipe trajectory line.
Note: see chapter 2 to apply constraints.

FIGURE 3.32 Sketch Flange

The centers of 2 circles should be **coincident** to the end point of the pipe trajectory line.
Note: See Chapter 2 to apply constraints.

- Modify dimensions of 2 circles as shown in Figure 3.33 → Click OK to accept the sketch.

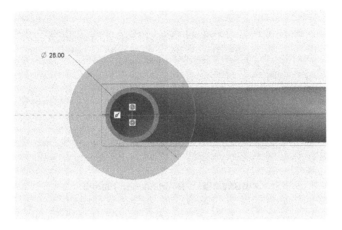

FIGURE 3.33   Enter Strong Dimension for Flange

- Select "Extrude" tool → Enter 2" depth → Click OK to accept the extrusion (Figure 3.34).

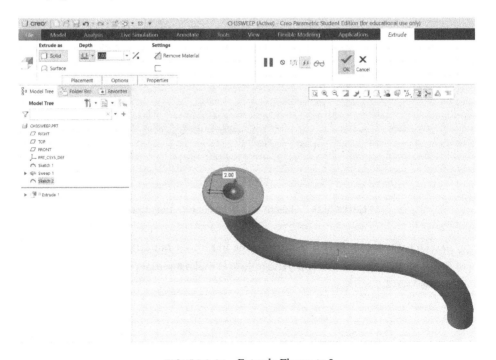

FIGURE 3.34   Extrude Flange to 2

- Repeat creating the lap flange at the other end of the pipe trajectory line to get the result as shown in Figure 3.35.

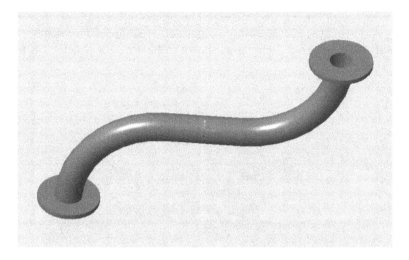

FIGURE 3.35 Build Second Flange

- Select the part (Figure 3.36) and select sketch.

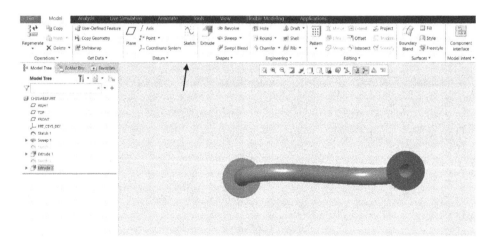

FIGURE 3.36 Sketching a Hole Feature to Pattern

- Sketch a circle (Figure 3.37) and extrude to remove material from flange.
- Extrude cut (remove material) the hole and make 8 radial patterns for both sides and save your work.

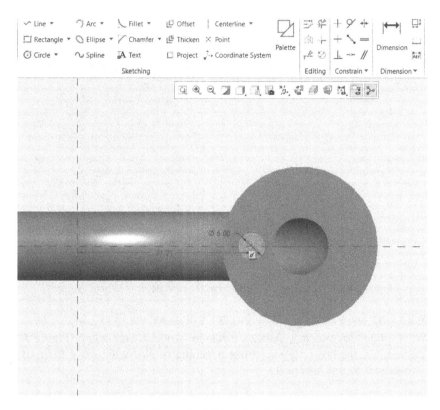

FIGURE 3.37   Removing Material to Build a Hole Feature

• Select Save 🖫 on left top corner or press Ctrl+S to save the part. Close the window by selecting ⊠ on the top left.

Note: Selecting ⊠ *on the top right will result in erasing current sessions and terminating Creo software completely.*

E3   **EXERCISE 3** | Steering Wheel (Combination of Revolve and Sweep)

• Select new working directory and start new part (see page 2).
• Name the part "steering-wheel."
   Note: We will choose inlbs_part_solid *as template.*

3.1.7 Start New Sketch

• Start Sketch and draw a geometry centerline vertically (Figure 3.38).

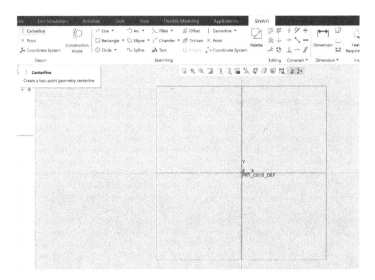

FIGURE 3.38 Starting a Revolved Feature for Steering Wheel Project

- With Line tool selected, Sketch the center hub as shown in Figure 3.39a.
- With Circle tool selected, Sketch a circle for wheel cross section (Figure 3.39b). Remember, we are only sketching a geometry. We will verify their driving dimensions and values next.

(a)

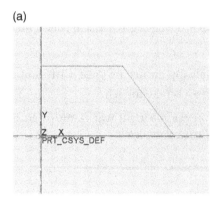

(b)

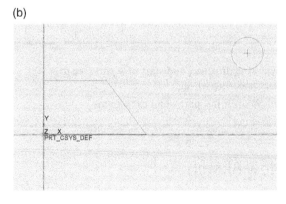

FIGURE 3.39 (a) Wheel Hub, (b) Wheel Handle Cross Section

Note: Press ESC to get back to Select mode and make sure the sketch is shaded.

- With Dimension tool, define driving (strong dimensions). Don't worry about the dimension values yet. (Figure 3.40).

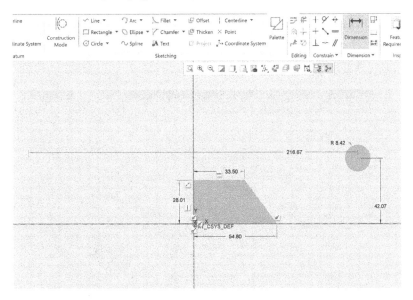

FIGURE 3.40 Define What Dimensions will Become Driving (Strong)

- With "Select" tool selected (you can press ESC twice) → Select all dimensions→Click "Modify."
- Insert the following dimensions (Figure 3.41) → Click OK to accept new dimensions (Figure 3.41) → Accept and exit Sketch mode.

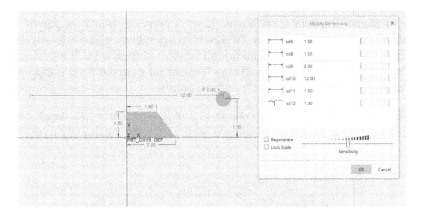

FIGURE 3.41 Entering Correct Values in Modify Dialog Box

3.1.8 Revolve Center Hub and Handle

- With the Revolve tool (Figure 3.42), verify the angle of revolution is **360.0** → Click OK to accept the revolved geometry (Figure 3.43).

Note: Ensure the sketched cross section is selected on the Model Tree to Revolve.

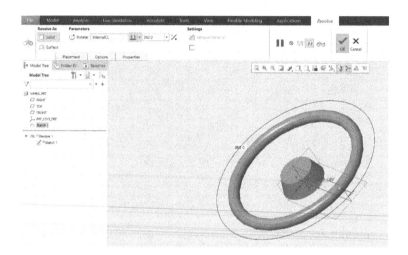

FIGURE 3.42 Revolve Selected Sketch

FIGURE 3.43 Verification of Angle of Revolution for Revolve

3.1.9 Central Hub Support Modeling

To develop supporting bars, we will need to work perpendicularly (or parallel to our initial sketch) to the steering wheel. To sketch on a perpendicular plane, select any of the intersecting datum planes. In addition to datum planes, we will introduce placing points with their coordinates and draw a spline for a curved handle support bars, which will go through the previously defined points.

- Select the vertical plane as shown in Figure 3.44 ➔ Enter the Sketch Mode ➔ Orient sketch parallel to your screen (Sketch View on the floating toolbar).

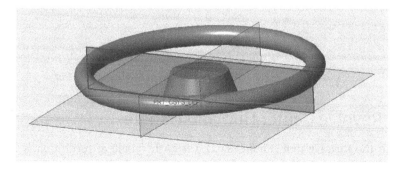

FIGURE 3.44 Selection of a Perpendicular Datum Plane for New Sketch

On the floating tool bar, Select "Display Type" (Figure 3.45) → Select "Wireframe" in the drop box (or press **Ctrl + 6**) to display the object in wireframe mode to see through the geometry.

Note: To orient the plane parallel to user interface, select Sketch View button.

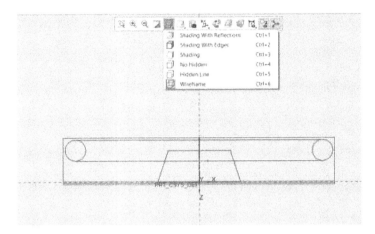

FIGURE 3.45   Changing Display Type on the Floating Toolbar (Graphics Toolbar)

- Select "Point" tool on the Ribbon (Figure 3.46a) → Place 4 points with the following coordinates: Point 1(Mid-Point of Hub), Point 2(X = 3.25, Y = 1.2), Point 3(X = 4.5, Y = 1.4), Point 4 (Center of Handlebar) as shown in Figure 3.46b.
- Define constraints and dimensions of 4 points as shown in Figure 3.46b.

(a)

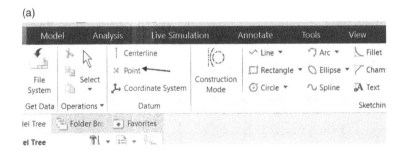

(b)

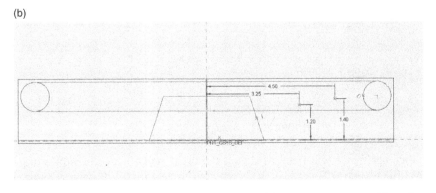

FIGURE 3.46   (a) Selecting a Point Tool on the Ribbon, (b) Entering Accurate Values for Each Point Coordinates

- Select "Spline" tool → Connect 4 points (Figure 3.47) → Press ESC → Click OK to accept the sketch & exit Sketch mode.

*Note: You might select "Shading" in Display Style drop box in floating toolbox (or press **Ctrl** + 3) to display Shaded model as default.*

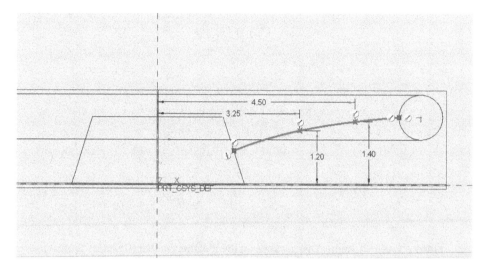

FIGURE 3.47   Connect All Four Points with a Spline Tool

3.1.10   Sweep Support Bar

- Select "Sweep" tool (Figure 3.48) → Ensure the trajectory line (spline) is selected.

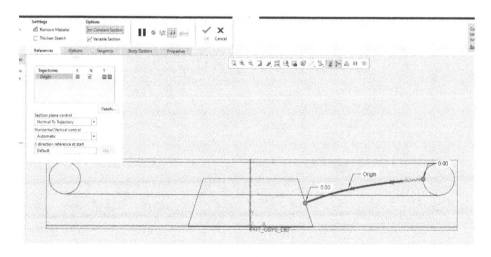

FIGURE 3.48   Sweeping Support Bars

- Once again, we will need to sketch the cross section for the supporting bars. Within Sweep → Enter Sketch → Sketch a circle with 0.5 diameter (Figure 3.49) → Click OK to accept sketch (Figure 3.50).

Note: You can change display type from the floating toolbar for convenience.

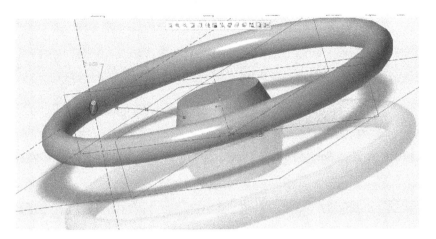

FIGURE 3.49   Sketching a Cross Section Within Sweep

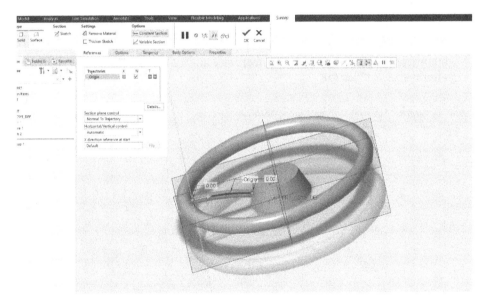

FIGURE 3.50   Select Options (Next to References) to Merge Ends

- Verify the sweep is working properly. You will notice the ends of the handlebar support do not snap to the round surfaces of the Hub and the Handlebar.
- Click Options (Still in the Sweep window next to References, Figure 3.50) → Check Merge Ends. Notice how the ends are merged → Click OK to accept changes.

3.1.11 Radial Pattern Handlebar Supports

- Select "Pattern" tool (Figure 3.51).

  Note: If no geometry is selected, the Pattern tool will be inactive.

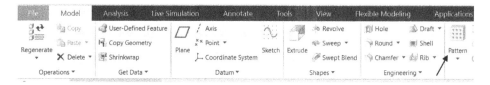

FIGURE 3.51 Select "Pattern" Tool on the Ribbon

- In Pattern, click Select Pattern Type drop box → Select "Axis" (Figure 3.52).

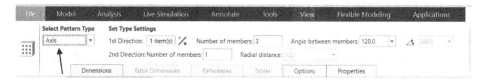

FIGURE 3.52 Select "Axis" Pattern Type for Radial pattern

- Select Revolve axis as Pattern axis (Figure 3.53) → Enter value of Number of members and Angle between members as 3 and 120.0, respectively, as shown in Figure 3.53 → Click OK.

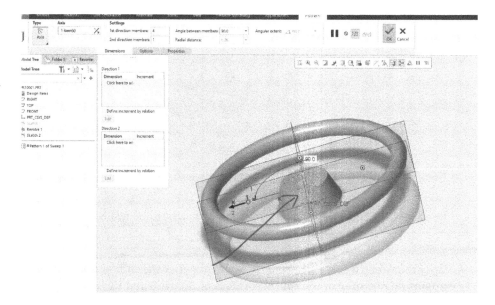

FIGURE 3.53 Select Geometry Axis to Pattern 3 Times Around the Selected Axis

- This will complete Sweep, Revolve, Pattern, Point, Spline Tutorials (Figure 3.54).

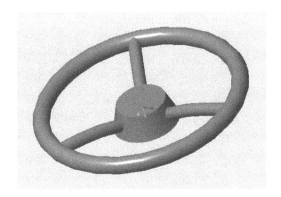

FIGURE 3.54   Final Steering Wheel Model

- Select Save on left top corner or press Ctrl+S to save the part. Close the window by selecting ⊠ on the top left.

  Note: Selecting ⊠ on the top right will result in erasing current sessions and terminating Creo software completely.

Chapter Problems

P3.1 Develop hub pulley with revolve command (metric) (Figure 3.55).

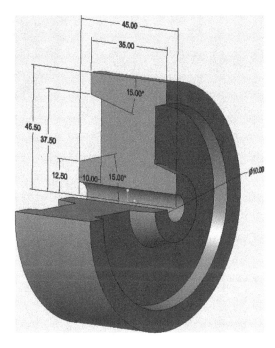

FIGURE 3.55   Hub Pulley

P3.2 Model friction wheel in metric units (Figure 3.56).

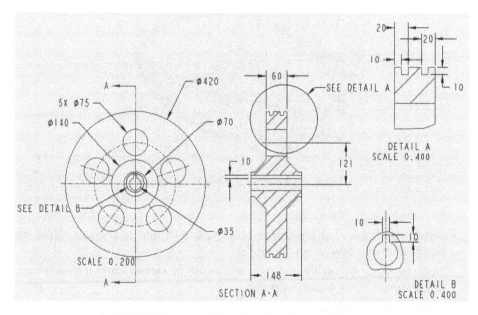

FIGURE 3.56   Friction Wheel Orthogonal View

P3.3 Develop a flanged pipe corner with sweep in metric units (Figure 3.57).

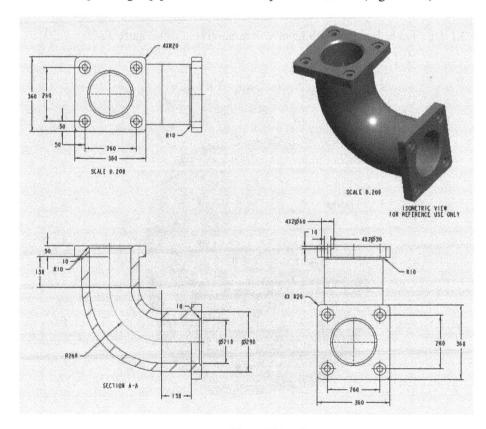

FIGURE 3.57   Flanged Pipe Corner

P3.4 Model connection block 166 (Figure 3.58).

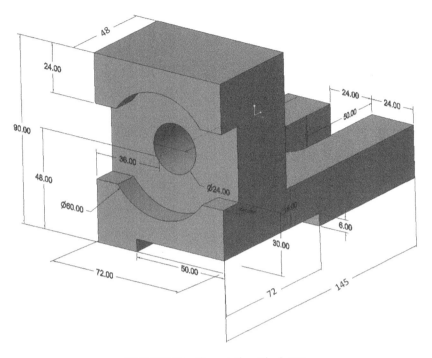

FIGURE 3.58   Connection Block 166

CHAPTER 4

3D Features/Modifiers

4.1 Introduction

Individual elements such as line, planes, and edges of a circle in two-dimensional sketch comprise a three-dimensional (3D) object or part as discussed in Chapter 2. In this chapter, we will learn how to develop a 3D part and its features with 3D tools such as directional and radial patterns, as well as hole tools.

CHAPTER OBJECTIVES

- 3D model development
- 3D tools and constraints
- Radial and directional patterns

Creo Parametric Modeling with Augmented Reality, First Edition. Ulan Dakeev.
© 2023 John Wiley & Sons, Inc. Published 2023 by John Wiley & Sons, Inc.

E1 **EXERCISE 1** | Radial Pattern (Bearing)

- Set up a new working directory and start a new part (Chapter 2).
- Name the part "Bearing." (Figure 4.1)

FIGURE 4.1 Bearing_231

Note: We will choose inlbs_part_solid as template.

- Click "Sketch" tool from the ribbon → Select "Front" plane → Click "Sketch" button on the dialog box.
- On the Display Toolbar, Click "Sketch View" button to orient the sketching plane parallel to the screen.
- Select "Circle" tool from the ribbon and draw a circle. (*Note: You can use revolve tool as well*).
- Hit escape and change the dimension to 140-in. diameter. (Figure 4.2).

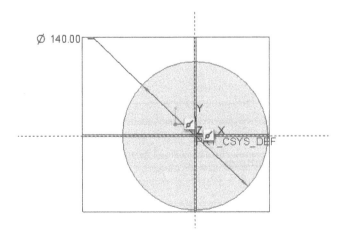

FIGURE 4.2 Circular Sketch to 140 Units

- Click "OK" to exit the Sketch mode.

E1   **EXERCISE 1** I Radial Pattern (Bearing) *(continued)*

- Extrude your circular sketch to 20 in. ➔ Click OK. (Figure 4.3)

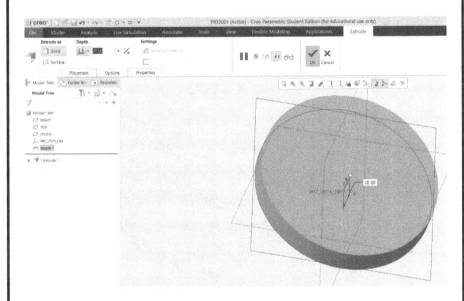

FIGURE 4.3   20-in. Extrusion

- Select one of the circle's flat surfaces and select "Sketch" tool to make a new 70-in. diameter circle (Figure 4.4).

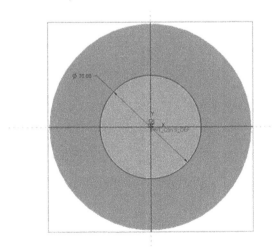

FIGURE 4.4   Sketch a New 70-in. Circle on Top of the Previous Geometry

- Verify the sketch dimension is strong and the circle is shaded ➔ Exit Sketch Mode.
- Extrude the last sketched circle to 34 in. (Figure 4.5) ➔ Click OK to accept the geometry.

(continued)

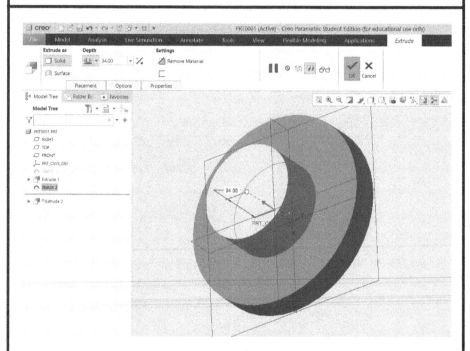

FIGURE 4.5 Extrusion of 70-in. Circle to 34 in.

- To make a through hole → Select the top surface of the model and Click "Sketch."
- Draw a 38-in. diameter circle (Figure 4.6) → Click OK to accept.

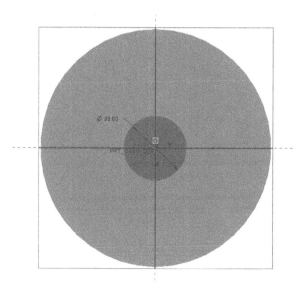

FIGURE 4.6 Sketch a 38-in. Circle to Extrude Cut the Hole

- Click the Extrude tool and change the direction to remove material (Figure 4.7).

E1 **EXERCISE 1** | Radial Pattern (Bearing) *(continued)*

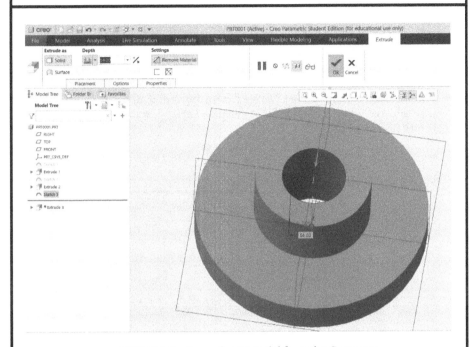

FIGURE 4.7 Removing Material from the Geometry

To introduce a radial pattern, first we will create a hole feature, select "Pattern" tool, and indicate "Radial" from options.

- Select the 140-in. surface ➜ Click "Sketch."
- Draw a 12-in. circle with 54 in. distance from the origin (Figure 4.8).

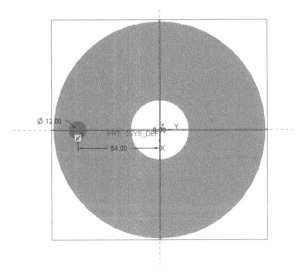

FIGURE 4.8 Draw a 12-in. Hole Feature for the Pattern

(continued)

E1   **EXERCISE 1** | Radial Pattern (Bearing) *(continued)*

- Click "OK" to accept the sketch → Extrude and remove material all the way to complete the hole feature.
- Select the hole (*Note: Not the sketch of the circle, but the entire hole feature*) → Click "Pattern" from the ribbon (Notice that the "Pattern" tool is hidden until the hole feature is selected).
- In the pattern menu → Click "Select Pattern Type" drop-down menu arrow and select "Axis" → In the "Number of members" enter 4 → Leave Angle at 90° for now (Figure 4.9).

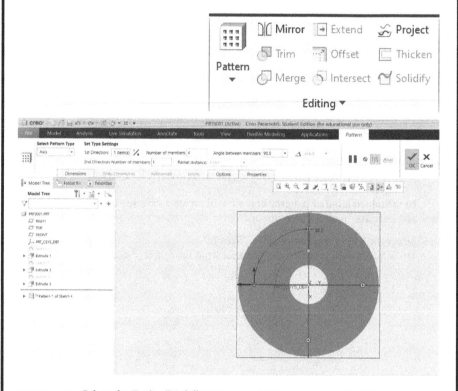

FIGURE 4.9   Select the Entire "Hole" Feature and Click Pattern to See Pattern Options

- Click "OK" to accept pattern (Figure 4.10).

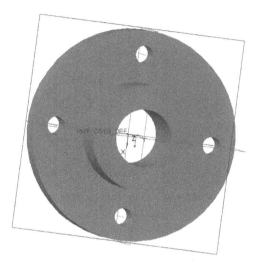

FIGURE 4.10   Complete Radial Pattern of the Hole Feature Around the Center Axis

- Select the edge of the circle as shown below (Figure 4.11) → Select "Round" tool on the ribbon→ Enter 2.5 in. → Click "OK" to complete the Bearing_231 Part (Figure 4.12).

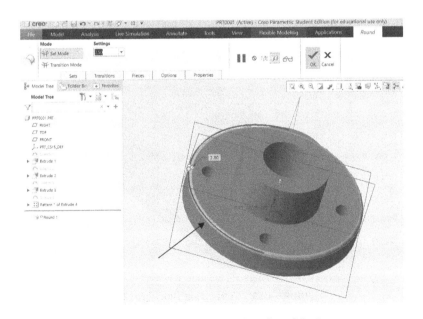

FIGURE 4.11   Add Round to the Edge of the Part

(*continued*)

E1 **EXERCISE 1** | Radial Pattern (Bearing) *(continued)*

FIGURE 4.12 Complete Bearing_231

- Save Bearing_231 part.

E2 **EXERCISE 2** | Directional Pattern (Lego-Type)

In this exercise, we will develop a block with features that follow a directional pattern.

- Start a new Part project (Figure 4.13).

FIGURE 4.13 Connection Block_20121

E2 **EXERCISE 2** | Directional Pattern (Lego-Type) *(continued)*

- Sketch Rectangle with the following dimension and extrude it to 9 in. (Figure 4.14).

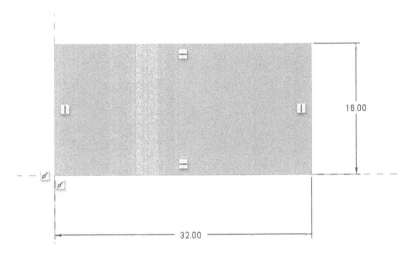

FIGURE 4.14 Connection Block_20121 Dimensions with 9-in. Thickness

- Select the top surface → Draw the following sketch (Figure 4.15) → Extrude to 1.5 in. → Click OK.

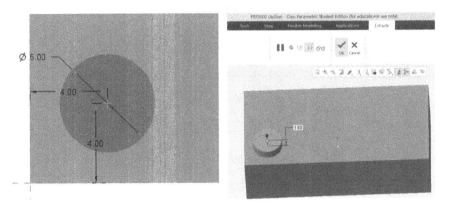

FIGURE 4.15 Part Feature Dimensions

- Select the new feature → Click Pattern tool → on "Select Pattern Type" select Direction → Select the Reference edge or surface (to which or from which the pattern direction should generate) → Define number of members (Figure 4.16) → Click OK (Figure 4.17).

  Note: You can repeat the same directional pattern for the second row. Additionally, you could model two features and pattern them together.

(continued)

E2 **EXERCISE 2** | Directional Pattern (Lego-Type) *(continued)*

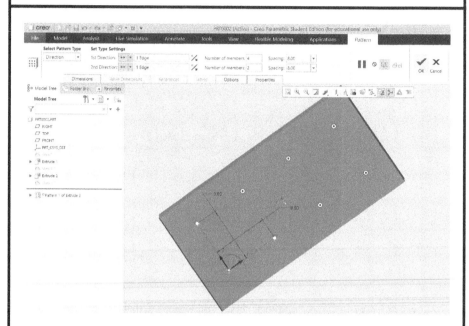

FIGURE 4.16 Directional Pattern Options

FIGURE 4.17 Complete Connection Block_20121 Part and Save in the Working Directory

E3   EXERCISE 3 | Helical Sweep (J Eye Hook Bolt)

- Start a new project (call it J_Hook_Bolt) with metric units (mmns_part_solid_abs) (Figure 4.18).

FIGURE 4.18   J Hook Bolt

- Using sweep method, sketch the model the following part (Figure 4.19).

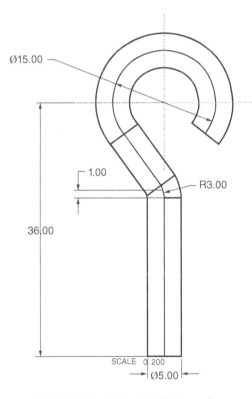

FIGURE 4.19   J Hook Bolt Dimensions

Sketching the Sweep

1. Sketch a 15-diameter circle with the origin with 36 mm origin from the bottom (Figure 4.20).

(*continued*)

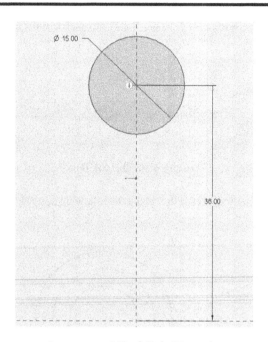

FIGURE 4.20 J Hook Bolt Dimensions

2. Sketch a vertical line from origin to the circle (Figure 4.21).

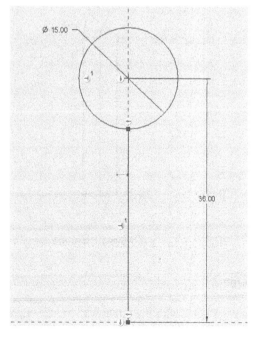

FIGURE 4.21 J Hook Bolt Dimensions

3. Sketch a randomly slanted line as shown in the figure (do not worry about the dimensions at this point) (Figure 4.22).

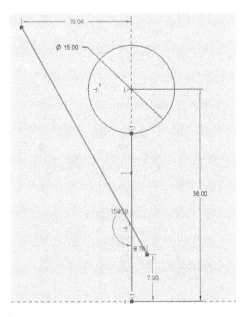

FIGURE 4.22 J Hook Bolt Dimensions

4. Select a tangent constraint ➜ Apply between the slanted line and the circle (Figure 4.23).

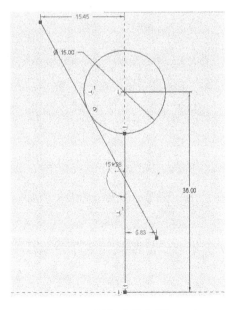

FIGURE 4.23 J Hook Bolt Dimensions

(continued)

E3 **EXERCISE 3** | Helical Sweep (J Eye Hook Bolt) *(continued)*

5. With the "Delete Segment" ➔ Trim the excess lines (Figure 4.24).

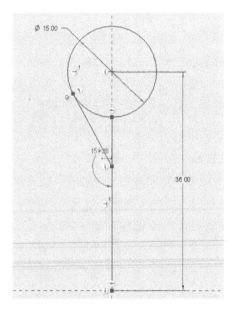

FIGURE 4.24 J Hook Bolt Dimensions

6. Select the "Fillet" tool ➔ Apply fillet between the slanted and the vertical lines ➔ Enter 3 for Radius and 1 for the distance (Figure 4.25).

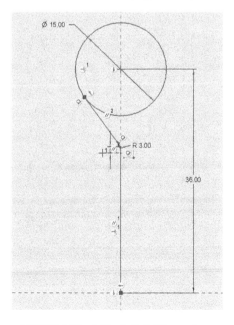

FIGURE 4.25 J Hook Bolt Dimensions

E3 | **EXERCISE 3** | Helical Sweep (J Eye Hook Bolt) *(continued)*

7. Sketch a 45-degree segment from the origin to the edge of the circle (Figure 4.26).

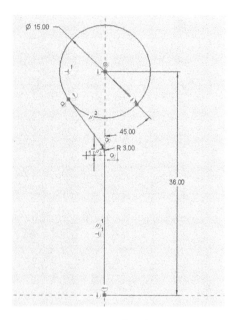

FIGURE 4.26 J Hook Bolt Dimensions

8. Trim the excess lines with Delete Segment tool (Notice a new weak dimension shows as we deleted the slanted 45-degree line) (Figure 4.27).

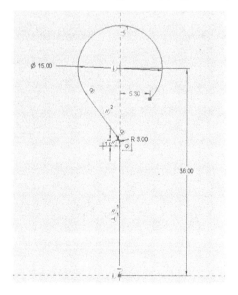

FIGURE 4.27 J Hook Bolt Dimensions

(continued)

E3   **EXERCISE 3** | Helical Sweep (J Eye Hook Bolt) *(continued)*

9. With "Horizontal" constraint tool → Select the point where circle and the line are targeted (P1) → Click the end point on the right (P2) (Figure 4.28).

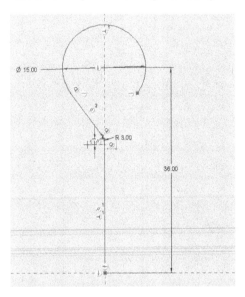

FIGURE 4.28   J Hook Bolt Dimensions

10. Click OK to exit Sketch.

11. Select Sweep from the Menu → Select the J Hook Sketch as a Trajectory Line (If the Arrow shows on the P2 point, single click on it so it shows at the bottom of the sketch, Figure 4.29) → Click Sketch.

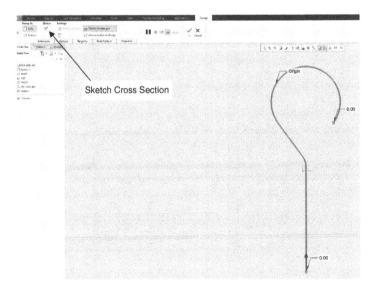

FIGURE 4.29   Sweep Section Sketch and Trajectory Line

E3 **EXERCISE 3** | Helical Sweep (J Eye Hook Bolt) *(continued)*

12. Draw a 5-diameter cross section at the intersection of two purple construction lines (*Note: If the sketch is not automatically oriented, click the "Sketch View" on the floating toolbar*) (Figure 4.30).

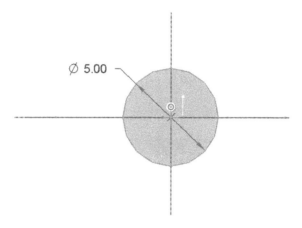

Ø 5.00

FIGURE 4.30 Cross Section Dimensions

13. To exit Sweep and review geometry (Figure 4.31).

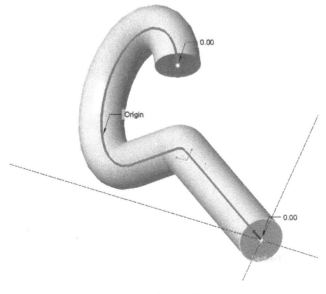

0.00

Origin

0.00

FIGURE 4.31 Review of the Geometry

(continued)

E3 **EXERCISE 3** | Helical Sweep (J Eye Hook Bolt) (*continued*)

Adding Threads with Helical Sweep

To add the threads of M10 bolt (you can add any threads based on the size of the part), we need to calculate the thread size with $Lg = 2d + 6$ for generic bolts (Figure 4.32).

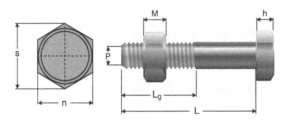

Gje ngebetegnelse. d	M6	M8	M10	M12	M16	M20	M24	M30	M36
Nøkkelvidde. n	10	13	17	19	24	30	36	46	55
Spissvidde. s	10.6	14.2	18.7	20.9	26.2	33	39.6	50.8	60.8
Hodehøyde. h	4	5.5	7	8	10	13	15	19	23
Mutterhøyde. m	5	6.5	8	10	13	16	19	24	29
Gje nget lengde. L_g $L_g = 2d + 6$ for L < 125 mm	18	22	26	30	38	46	54	66	78

FIGURE 4.32 Thread Length Table by Metric Unit Standard Bolts

Therefore, following the equation, we would get $Lg = 2 * 5 + 6 = 16$.

However, the J Hook bolt is not necessarily an M10 bolt for instance; therefore, our Lg would be: $Lg = 36$ (the length from the bottom to the origin) – Radius (15/2) – 1 for the fillet, which would become: $Lg = 36 - 15/2 - 1 = 15.5$ (very close to the standard calculator).

1. Orient the part Vertically ➔ Sketch a 15.5 mm vertical line from the bottom (*Note: You can select one of the reference planes to start your sketch*) ➔ Click OK to exit the Sketch Mode (Figure 4.33).

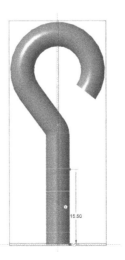

FIGURE 4.33 Thread Length

E3   **EXERCISE 3** | Helical Sweep (J Eye Hook Bolt) *(continued)*

2. Click the "Sweep" drop-down arrow on the menu → Select "Helical Sweep" (Figure 4.34).

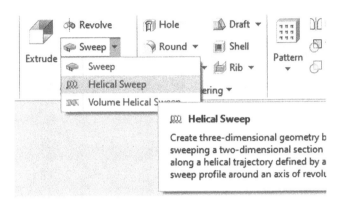

FIGURE 4.34   Helical Sweep Selection

3. Once in the Helical Sweep mode → Click "References" (It should be red) → Click Define for Helix Profile → Select the vertical line you sketched on step 1 → Click Define for the Hook Axis → Click Sketch (Figure 4.35).

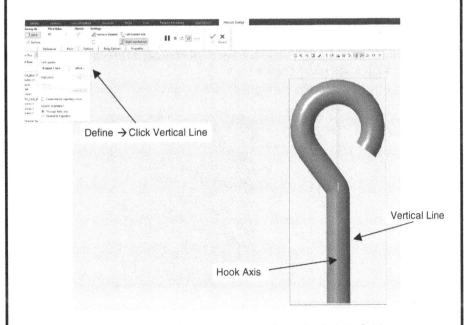

FIGURE 4.35   Helical Sweep Length and Revolve Axis Definition

(continued)

Now it is time to sketch the thread that will revolve cut around the hook axis to the vertical line height.

4. Click "Sketch View" on the floating toolbar to orient the part accurately.

5. Zoom to the bottom corner of the part where the two purple lines intersect each other (Figure 4.36).

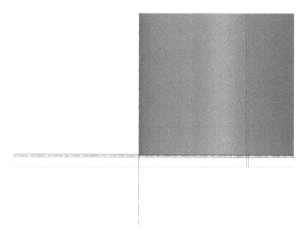

FIGURE 4.36 Thread Location (Intersection of Two Purple Construction Lines)

6. Sketch an equilateral triangle with 1.5 pitch (Remember, M10 bolt has 1.5 pitch) → Click OK to exit sketch (Figure 4.37).

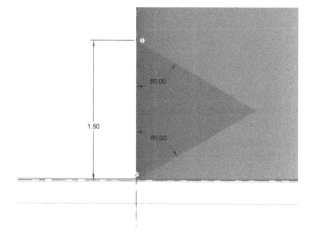

FIGURE 4.37 Sketching an M10 Bolt Size Threads

E3 **EXERCISE 3** | Helical Sweep (J Eye Hook Bolt) (*continued*)

7. Click "Remove Material" → Verify "Pitch Value" box is 1.5 → Click OK to accept (Figure 4.38).

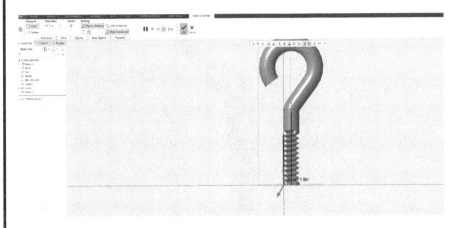

FIGURE 4.38 Remove Material, Verify Thread 1.50 Pitch

8. Select one of the vertical reference planes to sketch a triangular shape → Revolve cut the bottom (Figures 4.39 and 4.40).

(a)   (b)

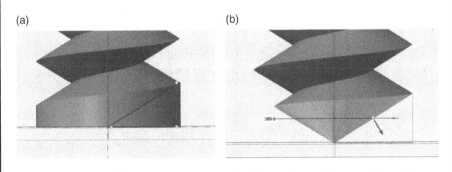

FIGURE 4.39 (a) Sketching a Pointy Tip, (b) Revolve and Remove Material

(*continued*)

E3 **EXERCISE 3** | Helical Sweep (J Eye Hook Bolt) (*continued*)

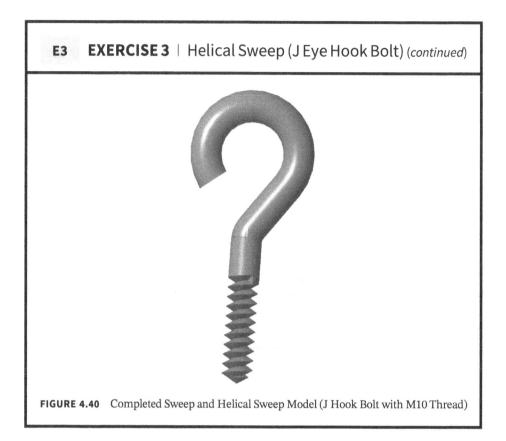

FIGURE 4.40 Completed Sweep and Helical Sweep Model (J Hook Bolt with M10 Thread)

Chapter Problems

P4.1 Model the following Bellbearing_1512 Part, add eight 0.12 diameter through hole with 1.125-in. distance from the center on the front flange (Figure 4.41).

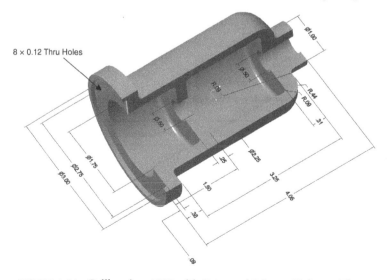

8 × 0.12 Thru Holes

FIGURE 4.41 Bellbearing_1512 with Patterned 8 Screw Holes on Flange

P4.2 Model Bearing_266 in metric units with 7 equally spaced 12 mm-diameter holes that are placed 52 mm from the part's center (Figure 4.42).

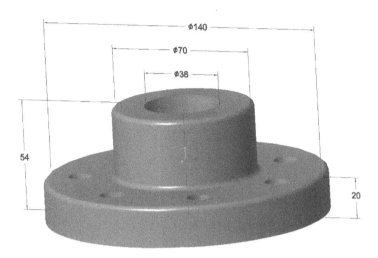

FIGURE 4.42 Bearing_266 (mmns)

P4.3 Model block 157 with directional pattern (Figure 4.43).

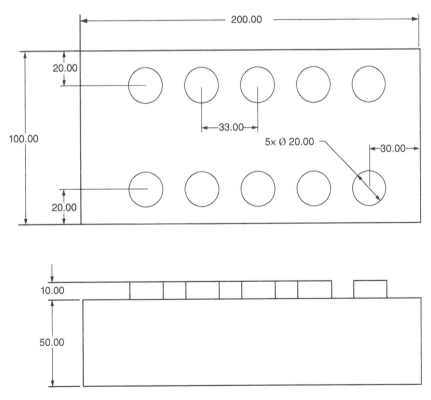

FIGURE 4.43 Block 157 (mmns)

P4.4 Model an M10 bolt using Figure 4.32 (Figure 4.44).

FIGURE 4.44   Block 157 (mmns)

P4.5 Develop an I-beam using dimensions below with imperial units (Figure 4.45).

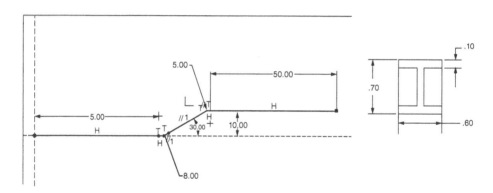

FIGURE 4.45   I-Beam

CHAPTER 5

Introduction to Drawing

5.1 Introduction

A drawing, or a technical drawing or an engineering drawing, is ultimately a communication tool that contains a rendered to a scale plan with detailed and precise dimensions as well as geometry shape of the manufactured product. In this chapter, we will start with the development of a basic drawing format that we will use throughout the drawing development.

CHAPTER OBJECTIVES

- Format development
- Generate orthogonal views
- Add dimensional and text annotations
- Datum axes and tolerances

> Note: This chapter shows how to generate drawings for parts. Assembly drawings with bill of materials and balloons will be discussed in the future.

5.1.1 Drawing Format/Template

When you work for an engineering firm as a design engineer, usually the firm provides their drawing template to place your modeled parts on. However, we will be generating our own format for this book and using it as a template. Per American Society of Mechanical Engineers (ASME) standards Y14.1-14.5 as shown in Figure 5.1, we will be generating a drawing format with a title block and revision block, and we will discuss the geometric dimensioning and tolerancing in future chapters.

- Start Creo Parametric (if it is not already running) → New → Format (Figure 5.2) → Change "frm0001" to "MyFormat" → Click OK.

Creo Parametric Modeling with Augmented Reality, First Edition. Ulan Dakeev.
© 2023 John Wiley & Sons, Inc. Published 2023 by John Wiley & Sons, Inc.

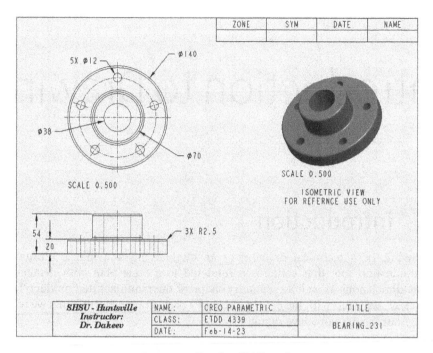

FIGURE 5.1 Bearing 231 Drawing

FIGURE 5.2 Select "Format" from New

Since majority of home or office printers use A size (or A4 in Europe) papers, will go ahead, and develop an A size format. On the "New Format" dialog box, select Standard Size A (A4 for Europe) from the drop down, verify it is landscape orientation, and click OK (Figure 5.3).

FIGURE 5.3   A Size Landscape Format

You will notice that the menu has changed slightly and the tools on the ribbon are different. First, we will develop a title block and place it at the bottom right corner of our format.

- Select "Table" menu ➔ Click "Table" drop-down arrow ➔ Select 3×3 Table ➔ Click to place on the workspace (Figure 5.4).

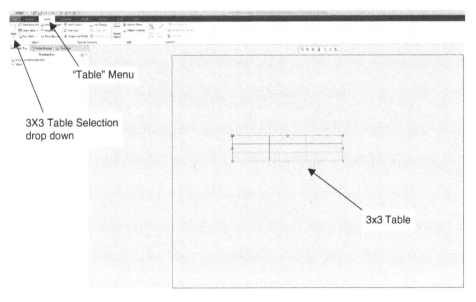

FIGURE 5.4   Place a 3×3 Table on the Work Area

Next, we are going to resize the columns. First column remains unchanged, second and third columns will be adjusted to 20- and 30-character sizes, respectively.

- Select any cell on the middle column (left click) ➔ Click right mouse button and Hold ➔ Select "Height and Width" ➔ Enter "20" in "Width" box ➔ Click OK (Figure 5.5).

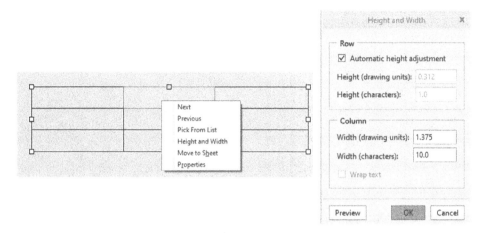

FIGURE 5.5 Adjust Width of the Column

- Repeat the previous step to adjust the third column's width to 30 characters.
- Select the Table ➔ Move your cursor to the Top Left of the Table (it will turn into 4 headed arrow) ➔ Click and drag it to place the table at the bottom right corner of the format (Figure 5.6).

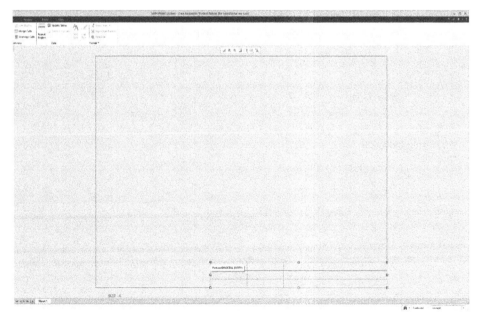

FIGURE 5.6 Move Table to the Bottom Right Corner

- Select two bottom cells (Ctrl+LMB) on the right column ➔ Click "Merge Cells" on the Ribbon (Figure 5.7).

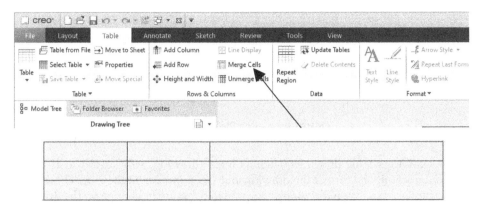

FIGURE 5.7   Merge Cells

- Insert a new 3×2 Table ➔ Merge all cells to make it one large cell/table ➔ Move it next to the previous table (Figure 5.8).

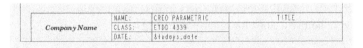

FIGURE 5.8   Insert New 3×2 Table

- At this time, we will enter text to make the Title block our own. You are welcome to change the text per your classes or company affiliations.
- Double click each cell to enter Text (Figure 5.8) ➔ When done entering text ➔ Click in the empty area to deselect table ➔ Single click the Box with "TITLE" ➔ Select "Format" on the Menu (Figure 5.9) ➔ Center.

FIGURE 5.9   Text Formatting

*Note: As you have noticed, we entered **&todays_date** text, this is a function Creo Parametric recognizes and automatically updates the actual drawing file creation date. Remember, we are developing a drawing format/template that we will use later to create drawings for parts. This function will update to a numerical value when we start a new drawing from this format.*

- Insert new 4×2 table for revision block ➔ Place it on top right corner ➔ Enter the following text (Figure 5.10).

Revision Block: A revision block is needed when a part's feature is updated. For instance, a car manufacturer decides to upgrade its car's headlights. This headlight might need a larger screw hole. Therefore, this update must be reflected on the

ZONE	SYM	DATE	NAME

FIGURE 5.10 Rev Block Placement

rev block. Larger scale drawings (such as E size drawing) will have A to H letters horizontally at the bottom of the drawing and 1–8 numbers on the side. These coordinates are used to locate the zone of the updated feature or change on the drawing. If the car's headlight bolt location is at A2, then the Revision block will contain A2. This will make the reviewing person to locate easier. Additionally, "SYM" column stands for "Symbol", which is used to indicate the change with a symbol of choice. You, as a design engineer, might have changed the headlight hole from M8 to M10 diameter and used A2 zone to locate and point the change (the hole) with H1 symbol. H-headlight, 1-revision 1. The date column will contain the date the drawing was revised, and the name will contain the person that reviewed and approved the change on the drawing before it goes into manufacturing.

- Click Save ➔ Ensure the type is .frm (Myformat.frm) ➔ Select the folder you will never forget ➔ Click "Save".
- Close the format (not the whole creo) ➔ Click "New" ➔ Drawing (Not Format) ➔ Uncheck "Use Default Format" (Figure 5.11) ➔ Click OK.

FIGURE 5.11 Start a New Drawing

Note: We unchecked "Use Default Template" because we want to use our newly generated format.

- On the new dialog box ➜ Select "Empty with format" ➜ Click "Browse" ➜ Find your saved format ➜ Click OK (Figure 5.12).

FIGURE 5.12   Empty with Format and Locate the Saved Format

- Verify that the date has been updated to today's date ➜ Close (no need to save) (Figure 5.13).

Company Name	NAME:	CREO PARAMETRIC	TITLE
	CLASS:	ETDD 4339	
	DATE:	Jun-01-22	

FIGURE 5.13   Updated Date

- Datum Axes and Tolerances.

E1   EXERCISE 1 ⏐ Three-Hole Brace (Part Modeling)

In this exercise, we will develop a model of three-hole brace part for exercise 2, which is the drawing development for the part.

- Select new working directory and start new part (Chapter 2: pg. 2).
- Name the part "Three_hole_brace".
  Note: we will choose inlbs_part_solid *to model the part.*

(*continued*)

E1	**EXERCISE 1** │ Three-Hole Brace (Part Modeling)

(continued)

- Click "Sketch" → Select "Front" plane → Click "Sketch" button on the dialog box.
- On the Display Toolbar (Floating Toolbar), click "Sketch View" to orient view parallel to the screen.
- Sketch a circle with 2.2-in. diameter (Figure 5.14).

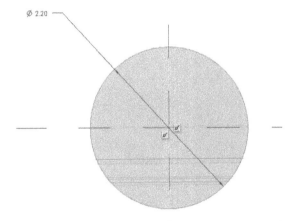

FIGURE 5.14 2.2-in. Diameter Circle

- Sketch another 1.5-in. diameter concentric circle (Figure 5.15).

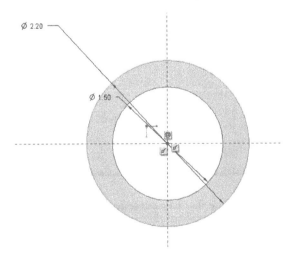

FIGURE 5.15 Sketching a Smaller Concentric Circle

- Now, you are going to sketch four more circles on both sides of these circles.
- Draw two 1-in. and 0.5-in. concentric circles on the right (Figure 5.16) with 2-inch distance from the origin.

E1	**EXERCISE 1**	Three-Hole Brace (Part Modeling)

(continued)

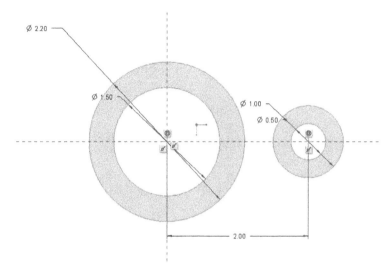

FIGURE 5.16 Concentric Circles

- Place a construction centerline vertically. This centerline will be used to mirror the two right circles to the left side (Figure 5.17).

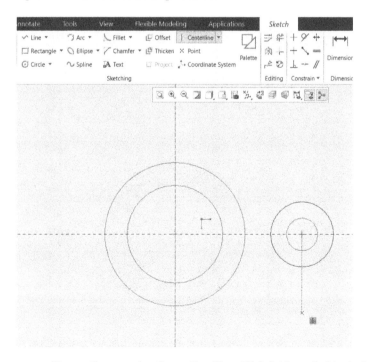

FIGURE 5.17 Place a Construction Centerline (Two Clicks) Along the Vertical Axis

(continued)

- Select two right circles → Select "Mirror" constraint → Click the construction centerline as a reference to mirror. The two circles are now placed on the left side of the origin (Figure 5.18).

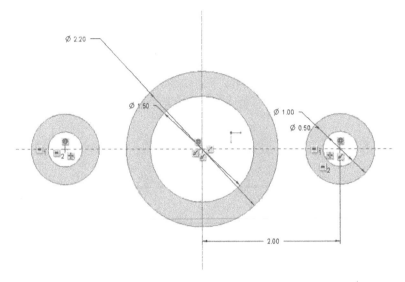

FIGURE 5.18   Mirrored Objects About the Construction Centerline Reference

To draw slanted lines and connect them to the respective circles, all we need to do is sketch some slanted lines, select "tangent" constraint, and select the objects (circles in this case) to connect to them.

- Draw a slanted line close to the outer circles (Figure 5.19). They do not need to touch as we are going to apply the "Tangent" constraint.

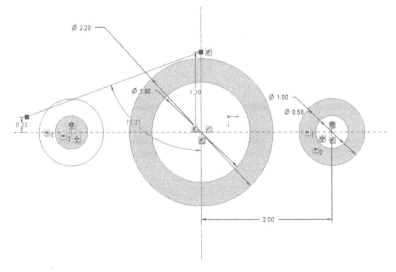

FIGURE 5.19   Sketch a Line Close to Outer Circles

- Click the "Tangent" constraint → Select the Line → Select outer circle to apply tangent.
- Repeat the previous step for the large circle in the middle (Figure 5.20).

  Note: Select the "Delete Segment" tool to trim excessive lines if necessary.

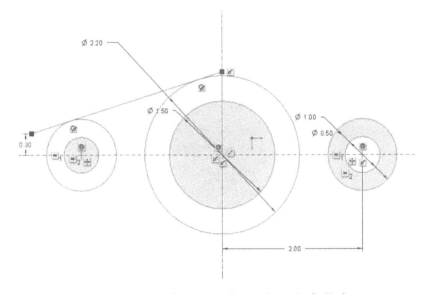

FIGURE 5.20   Apply Tangent Constraint to Both Circles

- Repeat the previous step three more times (Figure 5.21).
  Note: You can sketch slightly longer lines to ensure they are touching both circles as we can always trim them.

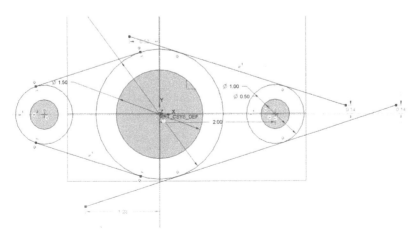

FIGURE 5.21   Apply Tangent Constraint Until All Lines Are Tangent with All Circles

(continued)

E1	**EXERCISE 1** ∣ Three-Hole Brace (Part Modeling)
	(continued)

- Trim all excessive lines with "Delete Segment" tool (Figure 5.22).

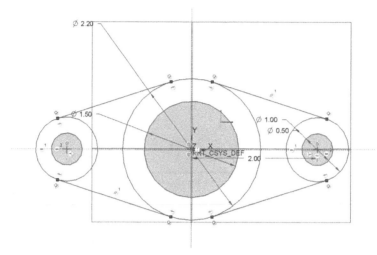

FIGURE 5.22   All Lines Are Tangent and Trimmed

Note that the sketch is not shaded as required. For a geometry to become "shaded" its outline must be connected with no intersecting lines. In this case, the circles are intersecting. Therefore, we need to trim circle segment to leave only the outer lines.

- With the "Delete Segment" tool, remove all the internal circle segments (Figure 5.23).

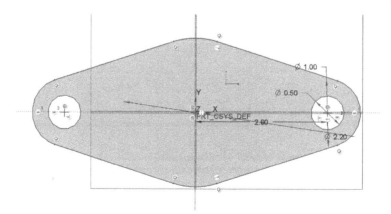

FIGURE 5.23   Trimmed Intersecting Segments

Now that the geometry is shaded with no weak (light blue color) dimensions, we can exit the sketch and extrude.

E1   **EXERCISE 1** | Three-Hole Brace (Part Modeling)
                          (*continued*)

- Click OK to accept Sketch.
- Extrude the sketch to 0.6-in. (Figure 5.24) ➔ Verify accuracy ➔ Click OK to accept Extrusion (Figure 5.25).

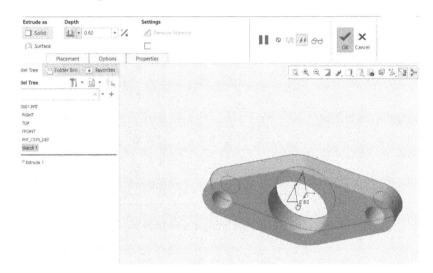

FIGURE 5.24   Verify Extrude Thickness Is 0.6 in.

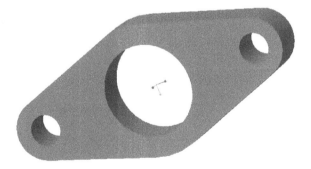

FIGURE 5.25   Completed Three_Hole_Brace Part

- Save the part.

E2   **EXERCISE 2** | Three-Hole Brace Drawing

- Start a new drawing ➔ Name "Three_Hole_Brace" ➔ Uncheck "Default Template" (because we will use our template) ➔ Click OK (Figure 5.26).

FIGURE 5.26   Defining a New Drawing

• On the "new drawing" dialog box ➔ Select "Empty with format" ➔ Click "Browse. . ." ➔ Find your saved template ➔ Click OK to accept (Figure 5.27).

FIGURE 5.27   Starting the Drawing with Saved Format

E2   EXERCISE 2 | Three-Hole Brace Drawing (*continued*)

Note that the new drawing is ready with our template and the date is current.

- Click "General View"  on the ribbon to insert active part → On the "Select Combine State" dialog box, select "Default all" → Click OK (Figure 5.28).

FIGURE 5.28   Default All to Insert Active Part as a Default Mode

- Now, clicking anywhere on the screen will insert a default view of the part → Select the top left corner of the drawing area to place the first default view (Figure 5.29).

FIGURE 5.29   Inserting the First View on Top Left Corner

Notice that the inserted view also automatically opens "Drawing Properties" and defaulted to "Drawing Type". Once you close this dialog box, you can always double click the view or click properties after right clicking on the view.

- On the Model View Names list, double click each view to see what views are available → Select the view that shows the top view of the part (Figure 5.30).

Note: Do not worry if your top view is named differently in the list. On this example, you can see the Top View on the drawing is derived from the "Front View" (Figure 5.30). We do not worry about the views because the view orientations can always be changed, additionally, the tool that manufactures the part will be referencing from the part itself. However, if your company wants you to keep the right orientation and start the sketch on the Front Plane, then you select the Front Datum during your sketch.

FIGURE 5.30 Finding Top View Orientation

- Select "View Display" Categorie ➔ Select "No Hidden" from the Display Style drop-down list ➔ Click Apply (Figure 5.31) ➔ Once satisfied, click OK to close the dialog box.

FIGURE 5.31 Apply "No Hidden" Style

Note: All orthogonal (2 dimensional) views must be presented with no shading, wither with or without hidden lines. Because the drawing is a communication tool, all the dimensions and notes will derive from respective orthogonal views. The only time we will keep the shading is on the Isometric View, which we will practice further ahead.

- On the floating graphics display toolbar ➔ Select "Datum Display Filters" ➔ Uncheck all (This will hide the datums from the drawing view, Figure 5.32).

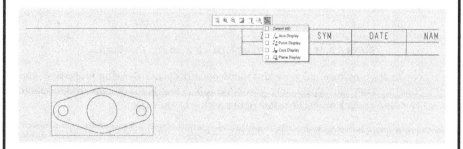

FIGURE 5.32 Hide All Datums from Drawing Views

Note: We will add axis to indicate the holes on the part; therefore, we will bring the axis datums, but those are part of the hole feature, not a reference datum.

- Select the Top View (a green box will appear around the view) ➔ Select "Annotate" tab on the menu bar ➔ Click "Show Model Annotations" (Figure 5.33). Notice that Creo Parametric provides this as a quick access when you select the view as well.

EXERCISE 2 | Three-Hole Brace Drawing (*continued*)

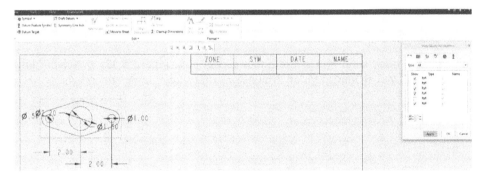

FIGURE 5.33 Show Model Annotations

Show Model Annotations: This tool (with the dialog box) will display all the dimensional values used to model the part. Additionally, the axis datums can also be brought with this tool. Initially, all the available dimensions are presented to the user to pick from. Once the driving (required) dimensions are selected, those dimensions will change color and show on the drawing view.

- Select the required driving dimensions by either clicking directly on the drawing or selecting the checkmarks in the dialog box (Figure 5.34).

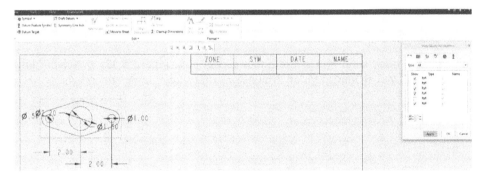

FIGURE 5.34 Select Driving Dimension to Keep on the View

- On the "Show model Annotations" dialog box ➔ Select Datum tab (first tab on the right) ➔ Select all axis datums to show on the view (Figure 5.35) ➔ Click OK to exit.

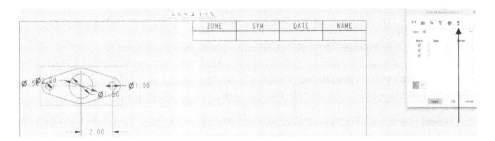

FIGURE 5.35 Select Axes for Each Hole of the Part

Now, we will clean up the top view and its dimensions:

- With your mouse pointer, select each dimension → Move to spread out (Figure 5.36).

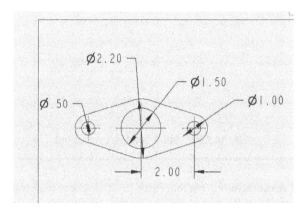

FIGURE 5.36   Moving Dimensions for Better Representation

Notice that the dimension leader arrows are double headed. These double-headed arrows display when we show hole's diameter, and a single-headed arrows display the radius of the arrow. However, some companies may use ONLY single-headed arrow to keep the view clean. It is also accurate as long as the dimension leader shows the "diameter" symbol. Next, we will change double-headed arrows to single-headed arrows:

- Select a dimension → On the quick access box (Figure 5.37) → Select "Flip Arrows" tool → Repeat for all dimensions that show diameter symbols.

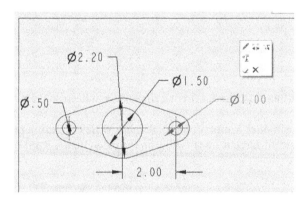

FIGURE 5.37   Flip Arrows for Diameter Dimensions

If your part is too close to the border and some of your dimensions do not fit, you can move the part to accommodate the annotations. To move the view:

- Select the view (Top View in this case) → Select "Layout" tab on the menu → Deactivate "Lock View Movement" on the ribbon (Figure 5.38) → Select the view (4-headed arrow appears) → Move the view to accommodate all annotations.

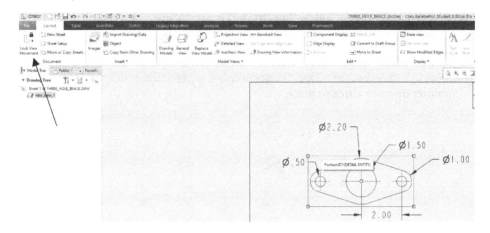

FIGURE 5.38   Move View to Accommodate All Annotations

Because there are two 0.5 diameter holes on both sides of the part, we will need to indicate it with 2X notation. To do so:

- Select the 0.5 diameter dimension (note ribbon tools change) → Select "Dimension Text" tool → Enter 2X in front of the @D (at dimension) symbol (Figure 5.39).

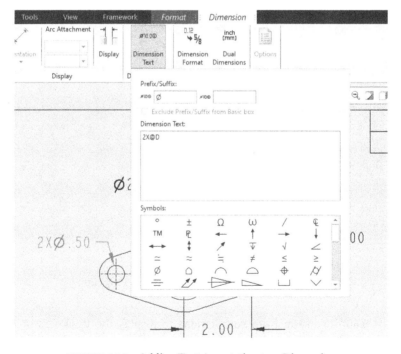

FIGURE 5.39   Adding Text Annotation to a Dimension

Note that the dimension on the Top View also updates as we enter the text.

At this point, we have declared all the driving dimensions, axes, and annotations on top view. To add more information such as thickness, we will need to show the side view. We could insert a new view from "General View"; however, it will not be connected to the top view, which we want. Therefore, we will be projecting the top view underneath:

- Select "Top View" → Click "Projection View" (Figure 5.40) → Move mouse pointer down → Click to place.

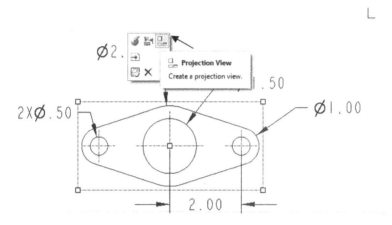

FIGURE 5.40   Projecting a New View from an Existing View

- Once projected (notice the shaded view) → Double click the projected view → View Display → Display Style (change to "No Hidden") → Click OK (Figure 5.41).

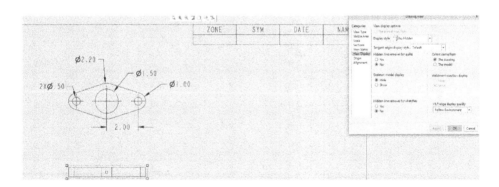

FIGURE 5.41   No Hidden Properties on Projection

- Insert 0.60 dimension from Annotations to display the thickness of the part (Figure 5.42).
- Insert a new "General View" on the Top right corner of the drawing sheet (This is a free-floating view because it is an isometric view that is used for reference only). → Under Annotate menu, click "Note" tool (Figure 5.42) → Place under the Isometric View (Shaded) → Type "ISOMETRIC VIEW FOR REFERENCE USE ONLY".

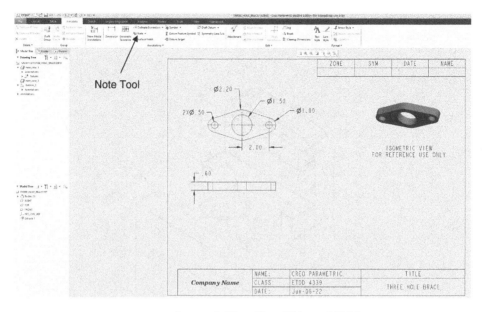

FIGURE 5.42   Isometric View, Note, Title, and Thickness

Chapter Problems

P5.1 Model Bearing_166a and develop a drawing using your template. All units are in millimeters (Figure 5.43).

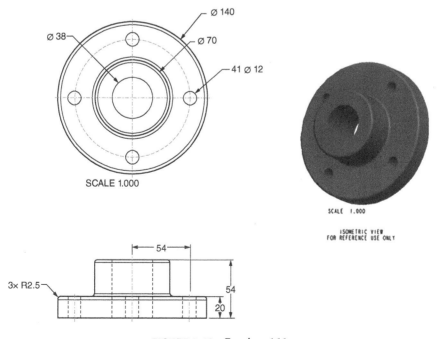

FIGURE 5.43   Bearing_166

P5.2 Model the Bracket 36622 and produce three orthogonal views (Top, Front, and Side) and isometric view on your drawing template. All units are in inches (Figure 5.44).

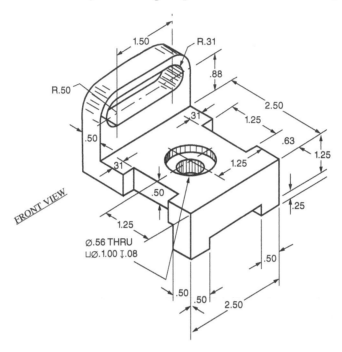

FIGURE 5.44 Bracket 366522

P5.3 Model and generate a drawing with all necessary orthogonal views to represent all given dimensions and the isometric view for reference on your template. All units are in millimeters (Figure 5.45).

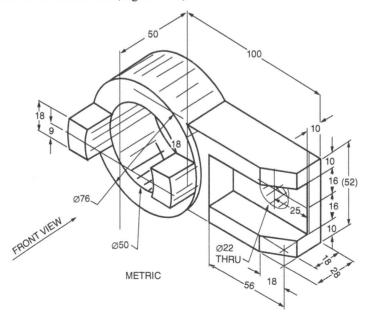

FIGURE 5.45 Lever Hub 420F

P5.4 Develop a drawing with all orthogonal views to represent all given dimensions and the isometric view for reference on your template for P4.2 problem on Chapter 4.

P5.5 Develop a drawing with all orthogonal views to represent all given dimensions and the isometric view for reference on your template for P4.3 problem on Chapter 4

CHAPTER 6

Drawing Tools

6.1 Introduction

In this chapter, we will discuss and develop drawing tools such as sections (planar and offset sections), auxiliary views, detailed views, dimension tolerances, orientation of the part, and radial datums. All these tools are necessary to communicate the design intent of the product further accurately to customers and suppliers.

Sections or cross sections are the cut-through views of parts to allow the person to see interior features of the modeled part, material, dimensions, and other elements such as hidden lines or datum axes. Creo parametric allows us to develop sections directly referenced on the part and bring onto the drawing view. In this chapter, we will cover both planar and offset sections.

CHAPTER OBJECTIVES

After completing this chapter, you should:

- Understand how to change view orientation.
- Learn tools within View Manager.
- Apply and understand planar and offset section views and add arrows.
- Add radial datum axis in detail options.
- Add detail views and dimensions.
- Add auxiliary views and dimensions.
- Learn Profile Rib feature.

Creo Parametric Modeling with Augmented Reality, First Edition. Ulan Dakeev.
© 2023 John Wiley & Sons, Inc. Published 2023 by John Wiley & Sons, Inc.

- Start Creo (if not already running) ➔ Setup Working Directory ➔ Model "Wheel" part (Figure 6.1) in inches ➔ Save.

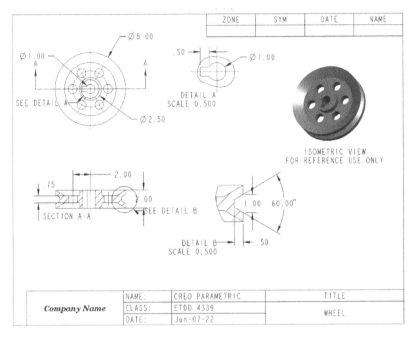

FIGURE 6.1 Planar Section and Detail View of the Part

- Start a new drawing (inches with your format) ➔ Call it "Wheel" (Figure 6.2).

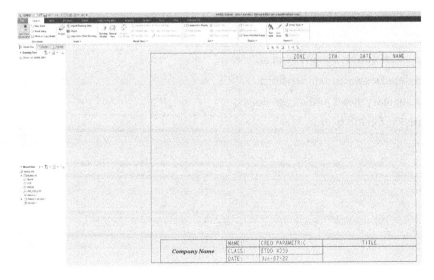

FIGURE 6.2 Wheel Drawing Start

E1   EXERCISE 1 | Planar Section (*continued*)

- Insert Top View on Top Left of the drawing → Change display to "No Hidden" →
  Hide all datums (floating toolbar, Figure 6.3).

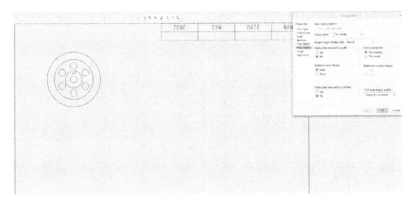

FIGURE 6.3   General Top View

Orientation: We will use the drawing "alignment" option to orient the view on the drawing. Later in the chapter, we will learn to orient the view in the part mode, save it, and bring to onto the drawing. Notice that the wheel part's keyway is oriented vertically pointing to the top, we will use "Drawing View (double click the view)" to change the orientation:

- Double click to open drawing properties (unless it is already open) → Select "View
  Type" on the list → Under the "View orientation" section, click "Angles" radio
  button → Keep "Rotation reference" Normal and type 90 in the "Angle value"
  (Figure 6.4) → Click "Apply" button.

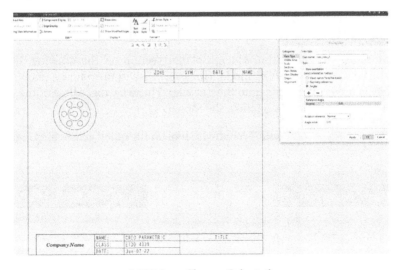

FIGURE 6.4   Change Orientation

Notice that the keyway has rotated to face the left. If your keyway is not oriented as shown in Figure 6.4, keep applying 90-degree rotation until your part is oriented.

(*continued*)

E1 **EXERCISE 1** | Planar Section *(continued)*

- Project the Top View underneath ➔ Apply "No Hidden" Display (Figure 6.5).

  Note: When applying section, always ensure that the view is "No Hidden". Section view should not contain any hidden lines.

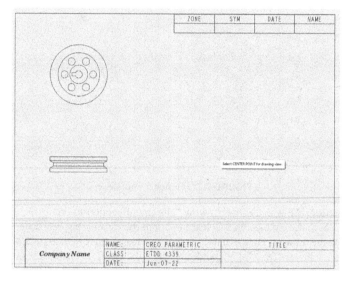

FIGURE 6.5 Projected Side View with "No Hidden" Display

6.1.1 Planar Section

There are multiple ways to introduce a planar section on the drawing, including section the view on the drawing mode. However, we will introduce a planar section as part of the model and bring onto the drawing. This way, regardless of the drawing update, the section will always follow the model.

- Switch to "Wheel.prt" with "Windows" tool on the quick access bar (Figure 6.6).

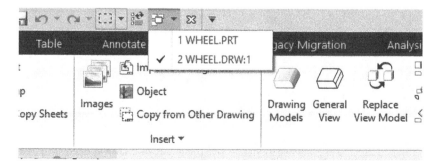

FIGURE 6.6 Switch Window from Drawing to Part Model

As you remember, we deactivated all datums on the floating toolbar to keep the drawing clean. Now, we will need to activate only the plane datums so we can introduce planar section on the datum plane.

- On the floating tool bar, activate only plane datums (Plane Display) ➔ Select the middle Datum Plane that cuts through the keyway (Figure 6.7) ➔ Select "View" Tab on the menu bar ➔ Under "Manage Views", select "View Manager".

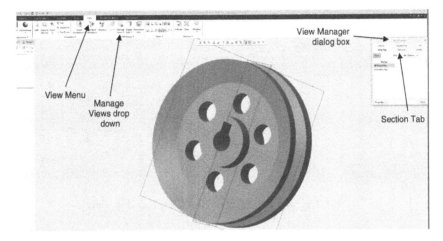

FIGURE 6.7   Select Middle Datum Plane, Activate View Manager

- On the View Manager, select "Sections" tab ➔ Click "New" drop down ➔ Select "Planar" on the list ➔ Rename the section name to "A" (Figure 6.8) ➔ Press Enter.

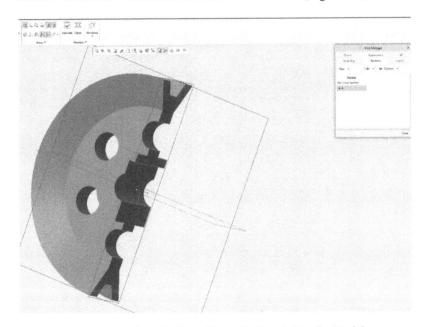

FIGURE 6.8   Introducing a Planar Section A-A to the Model

Note: If you wish to see the unsectioned whole part, double click "No Cross Section" on top of newly introduced section "A".

- Switch windows to drawing ➔ Hide Datum Planes (floating toolbar) ➔ Double click the projected side view (drawing properties will open) ➔ Select "Sections" category ➔ Select "2D cross-section" option (Figure 6.9) ➔ Click "+" button to add section ➔ Click Apply/OK.

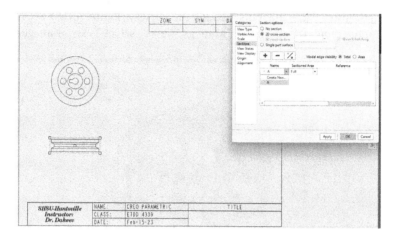

FIGURE 6.9 Applying Section A-A from Drawing View

You will notice that the projected view is sectioned, and the section name A-A is labeled under the projected view. It is important to add arrows to see the side of the section a person looking from. Therefore:

- Right click and Hold (right click mouse button) ➔ Select "Add Arrows" on the list ➔ Click the Top View (not the projected view) to add arrows (Figure 6.10).

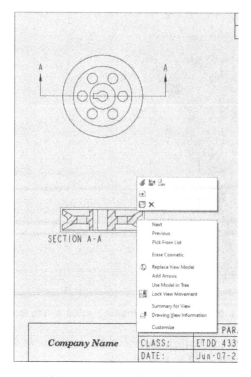

FIGURE 6.10 Add Arrows on Top View to Show Section View Side

- Add all dimensions and datum axes on both views as shown in Figure 6.11.

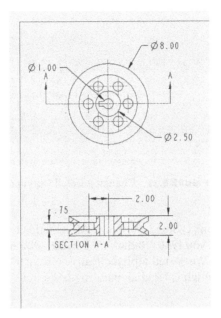

FIGURE 6.11   Dimensions and Datum Axes

6.1.2   Radial Datum Axis Circle

The top view contains the main axis on the origin and individual axes for each hole. Although it is correct and the projected view shows the distance between the center of one of the holes to the origin of the part, the axes do not represent the distance of all holes from the origin. Hence, we will need to introduce radial datum axis through drawing detail options:

- Select "File" menu ➔ Prepare ➔ Drawing Properties (Figure 6.12).

FIGURE 6.12   Drawing Properties

- On the Drawing Properties Dialog box → Click "Change" on Detail Options (Figure 6.13).

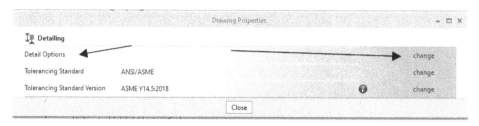

FIGURE 6.13 Changing Detail Options

- On the Detail Options Dialog box → Sort Alphabetically → Type "Radial" in the "Option" box → As you type "Radial. . ." a radial_dimension_display will highlight (Figure 6.14). We sorted alphabetically so we can quickly access the next option in the list, which is "Radial_pattern_axis_circle."

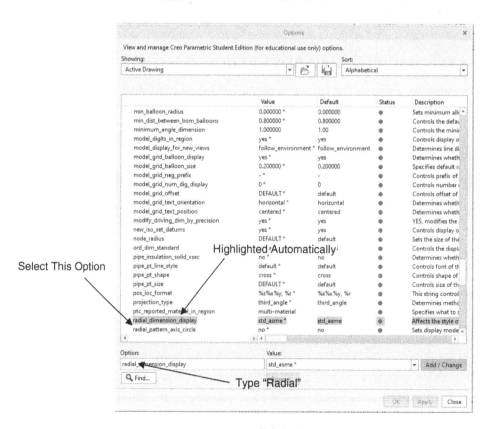

FIGURE 6.14 Radial Options

- Select "Radial_Pattern_Axis_Circle" option → Change "Value" from "No" to "Yes" (Figure 6.15) → Click "Add/Change" to accept "Yes" → Click OK to close Options → Click Close to close **Drawing Properties**.

FIGURE 6.15   Change Value from No to Yes

You may not directly see the changes we made in the drawing detail options.

- On the drawing, zoom in and out with your mouse scroll wheel to see the radial datum appear on Top view (Figure 6.16).

  Note: If you do not see the radial datum, verify: 1-Your holes are modeled and patterned radially, and 2-Datum axes are checked from Show Model Annotations.

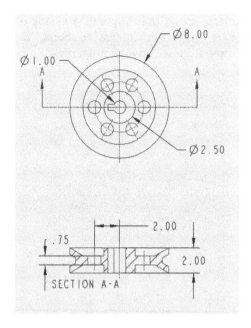

FIGURE 6.16 Radial Pattern Axis Circle Datum

6.1.3 Part Orientation

We changed the orientation of a view earlier. Additionally, we can save the orientation directly on the part. We will practice this on the isometric view:

- Bring a General View for Isometric (ensure it is shaded) → Add "Isometric View For Reference Only" note (Figure 6.17).

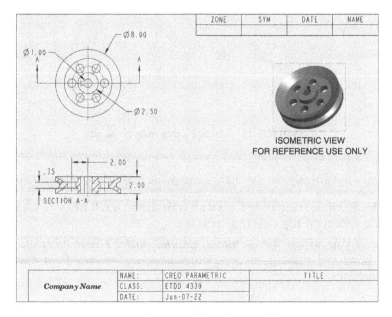

FIGURE 6.17 Isometric View Default Orientation

Notice that the default orientation may not represent majority of the features on the part. Therefore, we will change the part's orientation and call it on the drawing:

- Switch to Part Mode → Activate View Manager → Select "Orient" Tab (remember, before we used "Sections" tab → With your mouse wheel, rotate the part until the keyway points to the 7o'clock position with approximately 45° (you are eye-balling the orientation, Figure 6.18).
- Click "New" on the Orientation Tab → Type any name, we named it "Iso" for isometric (Figure 6.18) → Click Close.

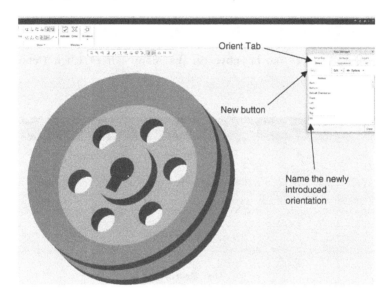

FIGURE 6.18 Saving New Orientation

- Switch back to Drawing mode → Double click Isometric View → Activate newly saved orientation from the list (Figure 6.19) → Verify the orientation is right → Click OK to accept.

FIGURE 6.19 Activating Saved Orientation

- Double click to enter the Title of the drawing under the "TITLE" on title block → Type "WHEEL" as the drawing title.

  Note: You may need to hold Alt button and double click the cell to enter text.

6.1.4 Detail View

Sometimes, to keep the drawing views clean and avoid congested dimensions, it is a good idea to introduce a detail view, which will not only enlarge the selected feature, but also keeps additional dimensions away from the view. To introduce a detailed view:

- Verify the "Layout" tab is active on the menu bar → Click Detailed View (Figure 6.20).

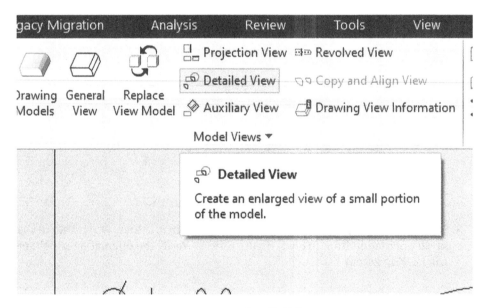

FIGURE 6.20 Detailed View

- Click the center of the feature (in this case the keyway) → Click around the center click to make a circular selection (Figure 6.21a).
- When finished, click middle mouse button (wheel) to complete selection (the selection will indicate with a circular sketch (Figure 6.21b)) → Left click on an empty spot to place a new detailed view (Figure 6.21c).

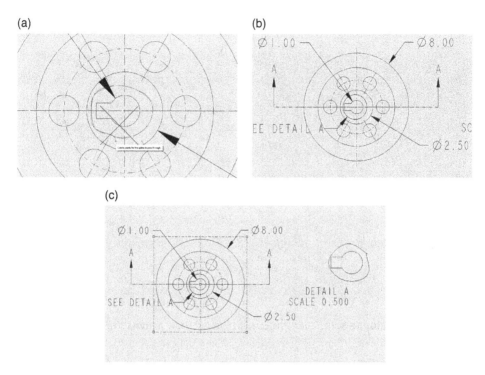

FIGURE 6.21 (a) Circular Selection of Detailed View (b) Detailed View Selection Complete and (c) Detailed View Placement

- Annotate the detailed view with the following dimensions and axis (Figure 6.22).

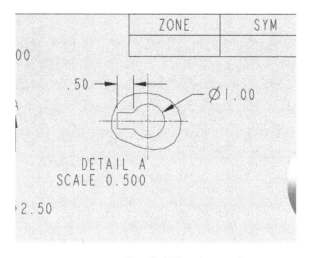

FIGURE 6.22 Detailed View Annotations

Note: If automatic show model annotations do not display all the dimensions, you can dimension them manually with the Dimension tool next to the "Show Model Annotation" tool under the "Annotate" tab on the menu bar.

- Repeat detailed view for the projected view with its dimensions (Figure 6.23).

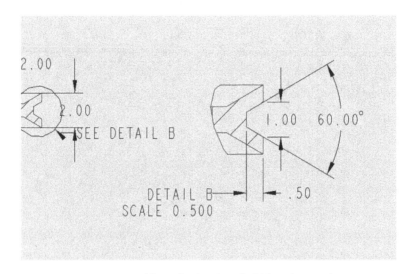

FIGURE 6.23 Side Projection Detailed View Annotations

- Save and close the drawing in your working directory (Figure 6.24).

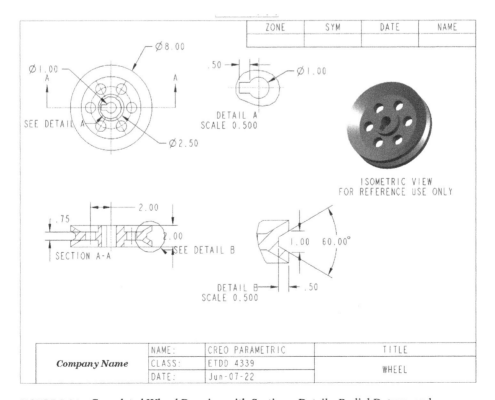

FIGURE 6.24 Completed Wheel Drawing with Sections, Details, Radial Datum, and Isometric Orientation

E2 **EXERCISE 2** | Offset Section

Offset section development is almost identical to the planar section, except we will have to sketch the section path instead of selecting a straight datum plane. For this exercise, we will model a bearing with three holes with counterbores and ribs (Figure 6.25).

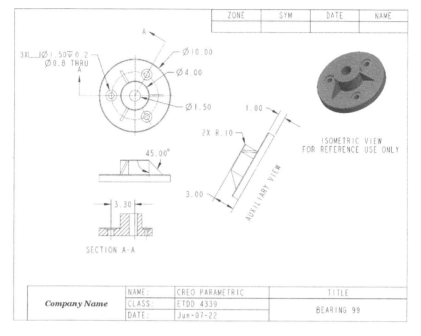

FIGURE 6.25 Offset Section Final Drawing

- Start Creo Parametric unless it is already running → Select a Working Directory → Model Bearing_99 in inches part solid units. You can revolve an L shape and add a 0.8 diameter through hole and a concentric 1.5 diameter circular cut that stops at 0.2 in. from the surface. You can pattern these two features together as a group until you have modeled majority of the part (Figure 6.26).
- We will start the rib feature next step:

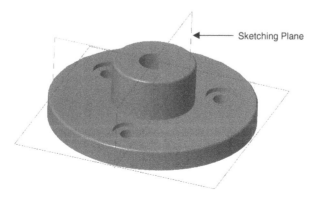

FIGURE 6.26 Bearing_99 with Counterbore Holes

(*continued*)

Sketching Profile Rib

- Select the plane that cuts through one of the holes (Figure 6.26) → Click Sketch → Orient the view parallel to your screen → Sketch a 45-degree line (Figure 6.27) → Click OK to exit Sketch.

FIGURE 6.27 Sketching Profile Rib

Note: Ensure you have sketched the line on the opposite side of holes. Otherwise, the rib feature will not generate on an empty spot created by holes.

- Select "Profile Rib" from the ribbon (Figure 6.28).

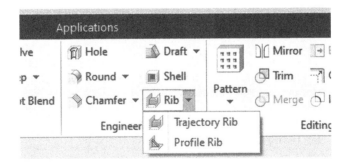

FIGURE 6.28 Select Profile Rib

- If the arrow shows outward → Click on the arrow to change its direction (Figure 6.29).

FIGURE 6.29 Click Arrow to Change Rib Direction

E2   **EXERCISE 2** | Offset Section (*continued*)

- Enter 0.3 as Rib thickness → Click OK to accept rib.
- Apply Radial pattern to complete 3 equally spaced ribs (Figure 6.30).

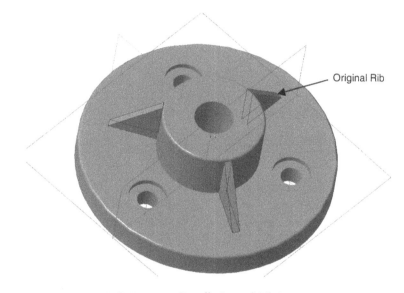

Original Rib

FIGURE 6.30   Equally Spaced Rib Features

6.1.5 Offset Section Drawing

- Start a new drawing with your format in inches → Place the Top View on top left corner.
- Apply "No Hidden" display → Hide all datums.
- Add only axis datums through "Show Model Annotations".
- Apply Radial Datum Circle (File → Prepare → Detail Options → Alphabetical Sort → "Radial. . ." → Add/Change → Yes.
- Orient as shown in Figure 6.31 (Holes on the left, Rib on the Right).
- Project a side view underneath → Apply "No Hidden" display → Switch to Part Mode.

Because we did not orient the top view earlier (unless you did), we do not know which hole is oriented to the left. However, we know which orientation gave us the top view (Figure 6.31). Therefore, we can activate the View Manager → From the list

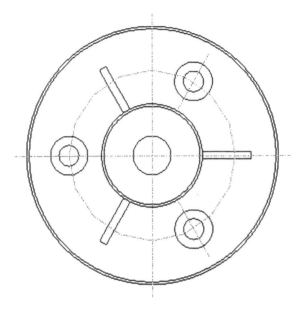

FIGURE 6.31   Radial Datums Applied and Oriented (Holes Left, Rib Right)

of orientations we can select the one on the drawing top view (or the one you oriented and saved, if you did) as shown in Figure 6.32.

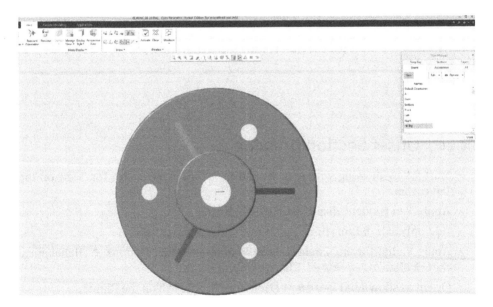

FIGURE 6.32   Finding the Top View Orientation in View Manager List

Note that the View Manager orientation list and the Drawing View Properties orientation list are same.

- Select the larger circle top surface → Sketch → Sketch a Line connecting the centers of Left → Center → Top Right Circles (Figure 6.33) → OK to accept Sketch.

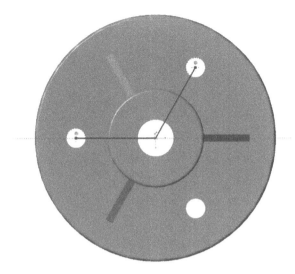

FIGURE 6.33   Sketch a Line Between Three Circles

- On View Manager → Section Tab → New → Offset (Figure 6.34) → Enter Section Name (A) → Select the sketched line (unless already selected) → Press Enter (Figure 6.35).

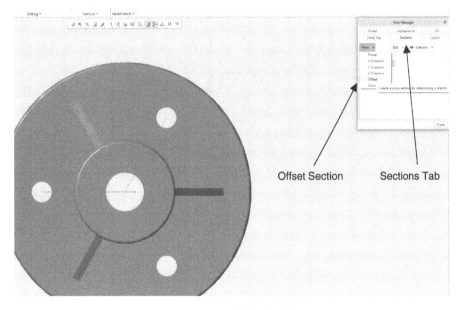

Offset Section   Sections Tab

FIGURE 6.34   Offset Section

- Click the section Arrow to flip the section view side (Figure 6.35).
- Click OK to accept Offset Section → Save your updated part → Change to Drawing window.

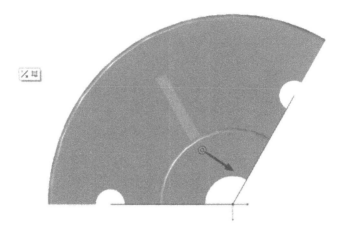

FIGURE 6.35 Offset Section on the Model

- Double click the projected view ➔ Sections ➔ 2D cross-section ➔ "+" ➔ A (unless you named differently) ➔ OK (Figure 6.36).

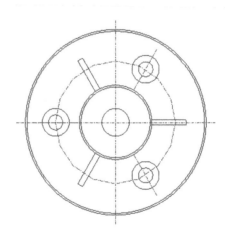

SECTION A-A

FIGURE 6.36 Offset Section

- Right click on the section view (hold right mouse button) ➔ Add Arrows ➔ Select the Top View to add section arrows ➔ Add Axes to the section view (Figure 6.37).

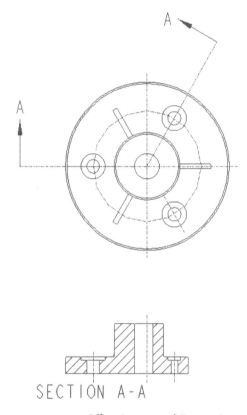

FIGURE 6.37   Offset Arrows and Datum Axes

Notice that the ribs are not shown on the Offset section since the offset line bypasses the rib on the right. To add dimensional annotations to the rib, you will need to project another side view.

6.1.6   Adding Text onto the Dimensional Values

To communicate counterbore and number of holes or ribs, and how deep do the counterbores go into the material, we will need to add text onto the dimensional annotations. We will practice two of these texts next:

- Add Show Model Annotations on the Top View ➔ Flip Arrows (Figure 6.38).

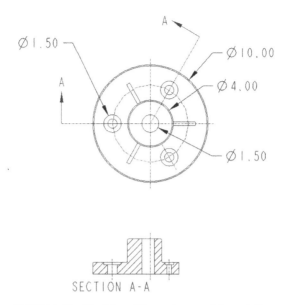

FIGURE 6.38 Top View Dimensions and Annotations

- Select 1.5 diameter dimension for three circles ➔ Click Dimension Text (Figure 6.39a) ➔ Type 3X in front of @D symbol ➔ Add counterbore symbol from the list ➔ Type the rest of the annotated text after @D symbol (Figure 6.39b).

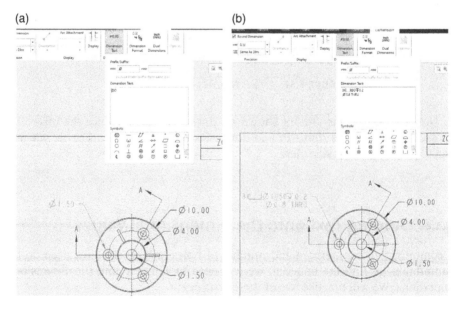

FIGURE 6.39 (a) Add Text to Dimension Value, (b) Added Text to Dimension Value

We read these annotations as "There are three 1.5 diameter counterbores (symbol) that stop at 0.2 depth (depth symbol)". The second line: "There is a 0.8 diameter through hole".

6.1.7   Adding an Auxiliary View

As discussed, the ribs are left behind the offset section, and we cannot see them on the side view. There are two options to add a side view: 1-Project another view (Figure 6.41) underneath (the problem with this is that we already projected that view for section and there is not much space for another view), and 2-Add an auxiliary view. The auxiliary view allows to pull a perpendicular view from an angled EDGE of the view. Here is how we add it:

- On "Layout" tab → Click "Auxiliary View" tool → Select the angled edge of the Profile Rib (Figure 6.40) → Drag mouse pointer on an empty spot → Click to Place.

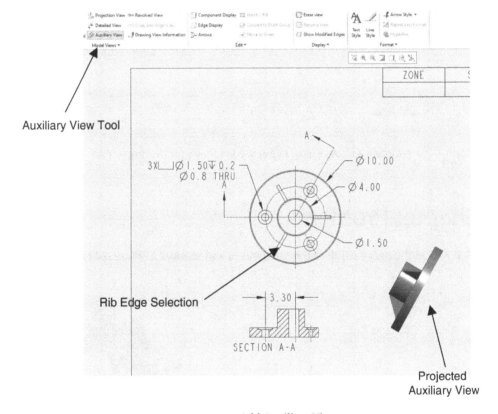

FIGURE 6.40   Add Auxiliary View

- Project a side view (Figure 6.41) → Add 45-degree angle.
- Add annotations for the Auxiliary View as shows on Figure 6.41.
- Add Isometric View and its note.
- Add Drawing Title on the Title Block.
- Save the drawing.

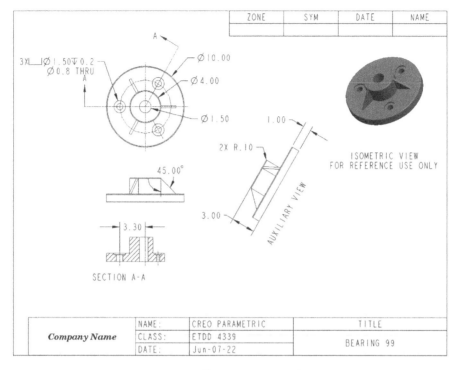

FIGURE 6.41   Auxiliary, Offset Section Views for Bearing 99 Part

Chapter Problems

P6.1 Model the part with its orthogonal views and sections (Figure 6.42).

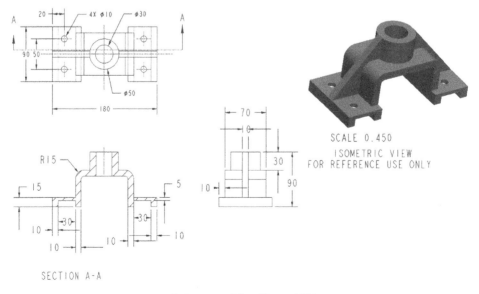

FIGURE 6.42   Vice Clamp 1523

P6.2 Model the part with its orthogonal views, auxiliary view, and sections (Figure 6.43).

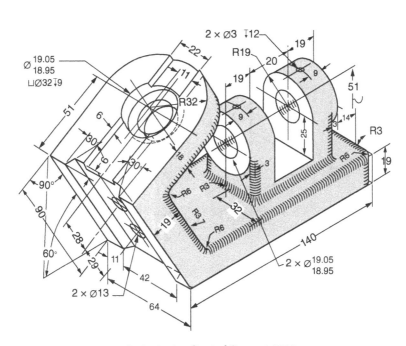

FIGURE 6.43 Control Support 1902

P6.3 Model the part with its orthogonal views, auxiliary view, and sections (Figure 6.44).

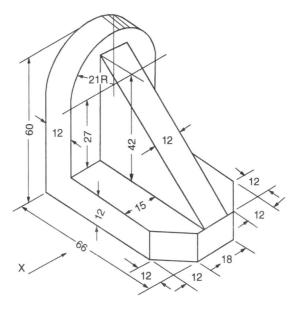

FIGURE 6.44 L-Stopper 944

P6.4 Model the part with its orthogonal views, auxiliary view, and sections (Figure 6.45).

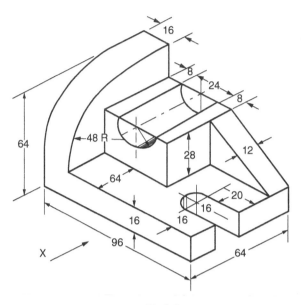

FIGURE 6.45 Shaft Support 10

CHAPTER 7

Assemblies

7.1   Introduction

Assembly of mechanical components is a capstone process in design and manufacturing. It brings together all previous chapters. However, in the industry, assemblies combine design, manufacturing, logistics, and communication to ensure the final product functions accurately. In this chapter, we will be developing some new parts to assemble them together; however, it is not likely that you will be developing from scratch in the industry. Often times, you update one or two components of a larger assembly, i.e. fender and its fastening components of a tractor without making changes to the rest of the tractor's parts. In this case, the assembly process requires the design engineer to open the larger assembly to visualize how the newly developed fender and its component fit or what design changes should the larger scale assembly requires to incorporate the new prototype.

CHAPTER OBJECTIVES

- Introduction to assembly
- Units of assemblies
- Assembly environment and tools
- Assembly constraint tools

E1   EXERCISE 1 | Swinging Link

This exercise will guide you step-by-step to finish assembly of a swinging link, including how to apply constraints and the relation of constraints.

(continued)

Creo Parametric Modeling with Augmented Reality, First Edition. Ulan Dakeev.
© 2023 John Wiley & Sons, Inc. Published 2023 by John Wiley & Sons, Inc.

E1   **EXERCISE 1** | Swinging Link *(continued)*

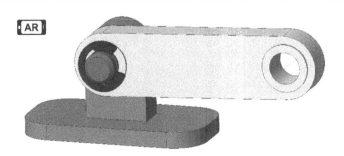

Select working directory *("Creo7_WorkingDirectory" or designated folder).*

➤ Start with modeling the following parts: *(can be found in Chapter 3 Problems)* (Figure 7.1)

• Create New Project (or *Ctrl + N*).

(a)

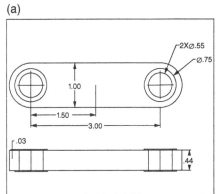

(b)

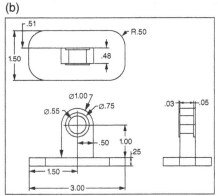

(c)

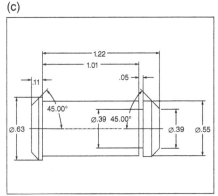

(d)

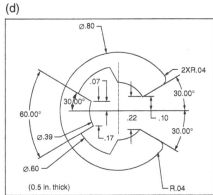

FIGURE 7.1   (a) Link, (b) Base, (c) Celvis Pin, (d) Retaining Ring

E1   **EXERCISE 1** | Swinging Link (*continued*)

As the dialog box appears, ensure that "Assembly" type is selected (Figure 7.2a).

- Enter the assembly name (you may call it "Swinging-Link") → Uncheck "Use default template" → Select OK (Figure 7.2a).
- Select "inlbs_asm_design_abs" as the unit (Figure 7.2b).

    Note: You may select different unit system based on how the parts are modeled.

(a)   (b)

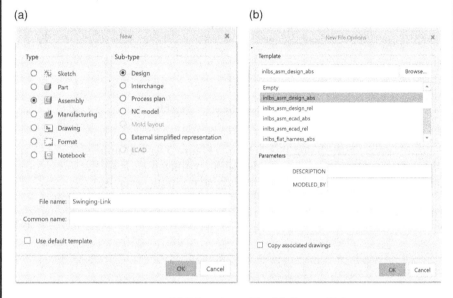

FIGURE 7.2   (a) Select Assembly, (b) Choose Units

7.1.1   Start Assembly

The Assembly session will appear in another window (Figure 7.3) after creating the assembly.

- On the Model tab, select Assemble to add a component to the assembly (Figure 7.3).

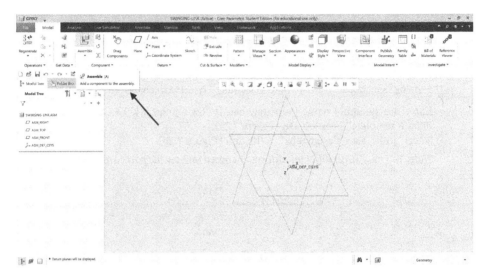

FIGURE 7.3 Select Assembly

From here, a new dialog box "Open" (Figure 7.4) will appear. As we set the working directory to destination folder with completed parts earlier, here you may choose the first part to assemble (which is base.prt in this exercise).

- Select "base.prt" → Select OK.

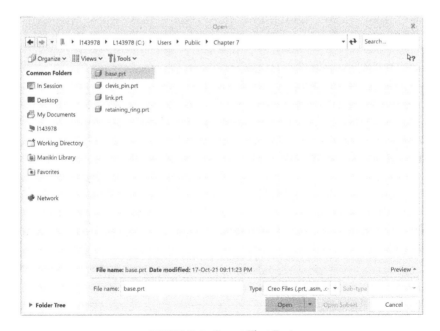

FIGURE 7.4 Insert First Part

Once you open the part, the Creo window will automatically switch to Component Placement tab (Figure 7.5). As the part is opened, you can use the middle mouse to rotate the part or Shift + Middle Mouse to move the part. Part orientation depends on how you modeled the parts previously.

- In the drop-down Constraints menu (Figure 7.5), choose Default → Select OK.

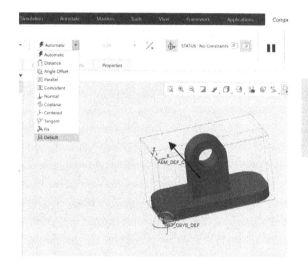

As the first part of the assembly will be assigned as coordination for the whole assembly, setting its constraint as default is important.

FIGURE 7.5   Default Constraints

As the first part of the assembly will be assigned as coordination for the whole assembly, setting its constraint as default is important.

After selecting Default, the part will turn orange color (Figure 7.6).

Note: Orange color identifies the part is Fully Defined.

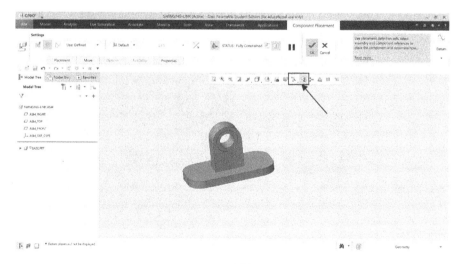

FIGURE 7.6   Hide Datums

- Deselect showing axes and planes for clearer view (Figure 7.6).
- Click OK to finish placing first part.

Repeat the previous step to bring the second part into the assembly.

- Select Assemble to add next component (Figure 7.7).

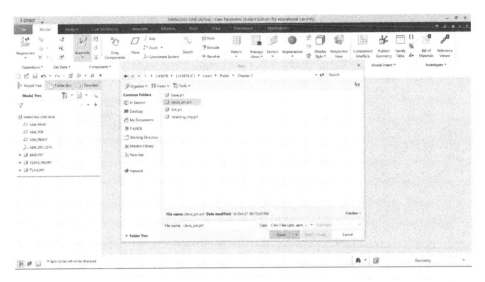

FIGURE 7.7 Insert Pin

- Select Clevis pin (clevis_pin.prt) → Select Open.

When the Component Placement tab is activated, in the constraints drop-down menu, select "Coincident" (Figure 7.8).

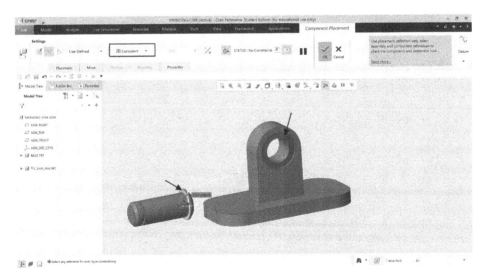

FIGURE 7.8 Coincident Constraint Flat

- As Coincident constraint is selected, you may need to select two features (Figure 7.8) that are coincident with each other – in this case, two planar surfaces. Constraint application will relocate the part according to the designated constraint (Figure 7.9).

- Continue to select the marked features to be coincident (Figure 7.9).

  Note: make sure to select "Conincident" constraints before selecting the features.

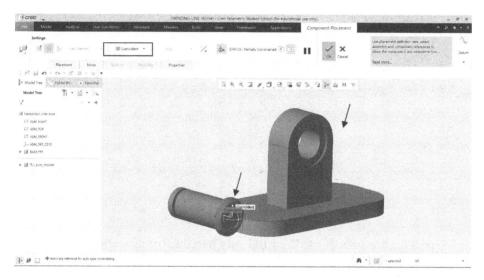

FIGURE 7.9 Coincident Constraint Ring

As the desire constraints are applied correctly, the part turns orange meaning fully constrained (Figure 7.10). In the placement tab, you can manage all the constraints, modify the applied constraints, and add new constraints (Figure 7.10).

FIGURE 7.10 Accept Constraints

- Make sure required constraints are applied and parts become fully constrained (Figure 7.10) → Select OK to return to view mode (Figure 7.11).

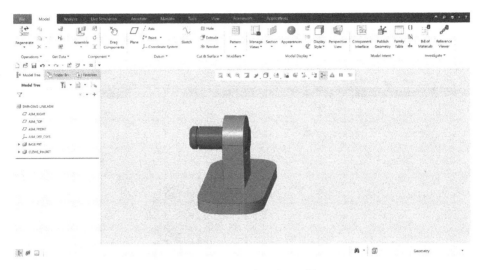

FIGURE 7.11 Review Assembly

Repeat the previous step to bring the third part into the assembly.

- Select Assemble to add next component (Figure 7.12).
- Select Link (link.prt) → Select Open.

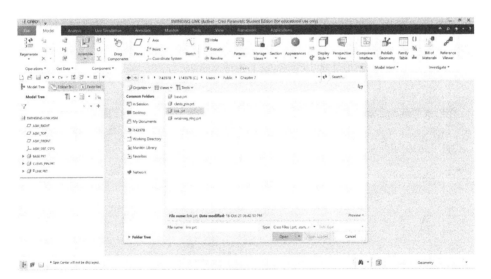

FIGURE 7.12 Insert Link

In Settings tab (Figure 7.13), select "Type of constraints" drop-down menu → Select Pin constraints.

> As some assemblies are required to be dynamic, pin constraint allow the components to spin or rotate about the axis of translation.

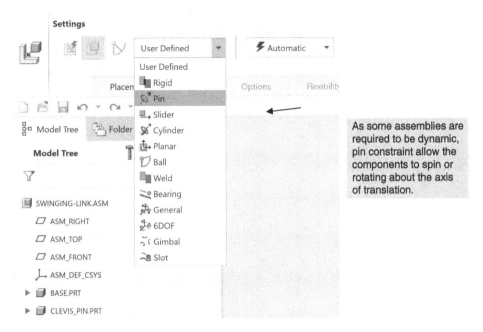

FIGURE 7.13   Add "Pin" Constraint

Open Placement tab (Figure 7.14) to manage constraint details.

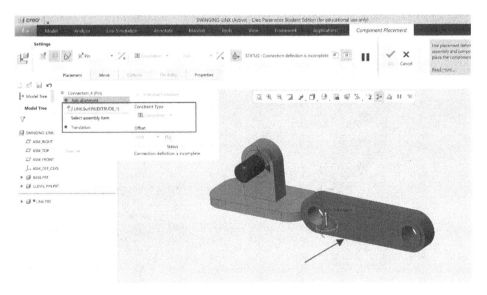

FIGURE 7.14   Select Round Surface

- Select Axis alignment (Figure 7.14) with "Coincident" as constraint type →
  Select 2 surfaces as shown in Figure 7.14 → the part will be automatically oriented as shown in Figure 7.15.

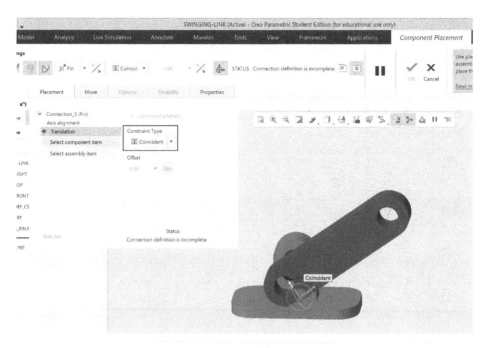

FIGURE 7.15 Orient Part

If the part is not oriented as Figure 7.15, make sure the constraint type is Coincident.

- Select Translation (Figure 7.15) with "Coincident" as constraint type → Select 2 surfaces as shown in Figure 7.16a,b

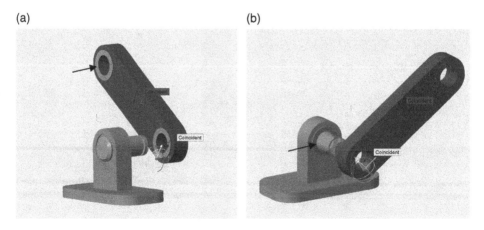

FIGURE 7.16 (a) Select Flat Surface for Link, (b) Select Flat Surface for Base

After setup constraints, the assembly will be automatically orientied and the part becomes orange, which identifies Connection Definition Complete (Figure 7.17) → Select OK to complete and return to view mode (Figure 7.18).

FIGURE 7.17   Verify Fully Constraint

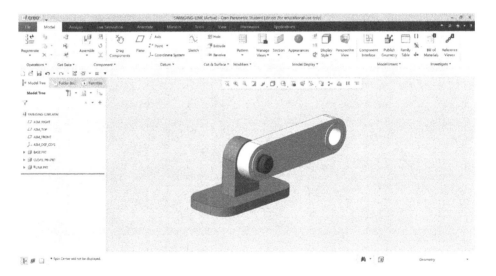

FIGURE 7.18   Review Assembly

Repeat the previous step to bring the last part into the assembly.

- Select Assemble to add next component (Figure 7.19).
- Select Retaining ring (retaining_ring.prt) → Select Open.

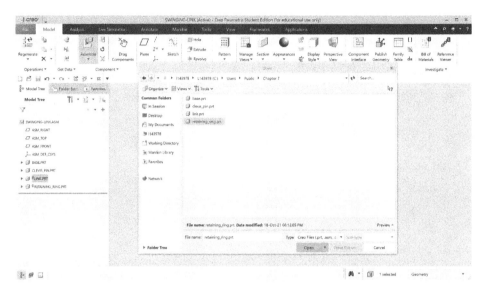

FIGURE 7.19   Insert Retaining Ring

In Settings tab (Figure 7.20), select "Type of constraints" drop-down menu →
Select "Pin" constraints.

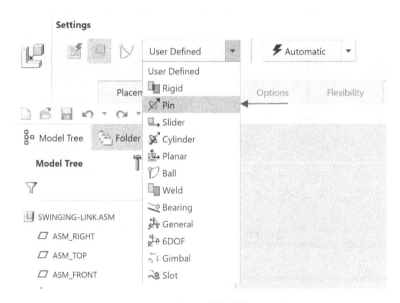

FIGURE 7.20   Insert "Pin" Constraint

Open Placement tab (Figure 7.21) to manage constraints easier.

• Select Axis alignment (Figure 7.21) with "Coincident" as constraint type →
Select 2 surfaces as shown in Figure 7.21 → the part will be automatically oriented as shown in Figure 7.22.

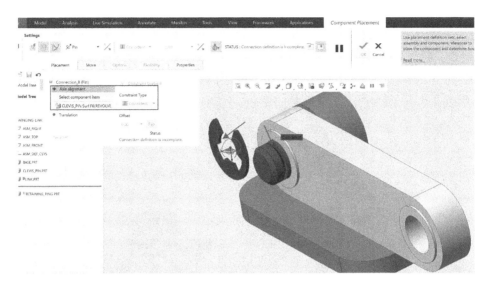

FIGURE 7.21   Select Round Surface

- Select Translation (Figure 7.22) with "Coincident" as constraint type.

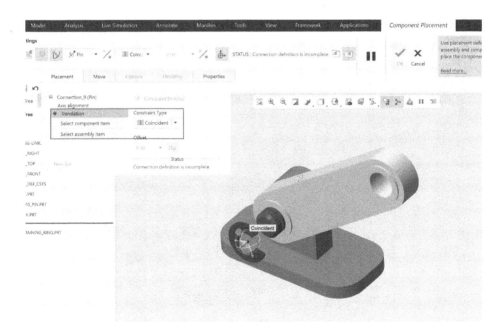

FIGURE 7.22   Coincident Constraint

- Select 2 surfaces as shown in Figure 7.23a,b.

(a) (b)

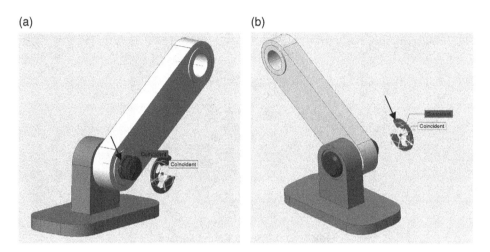

FIGURE 7.23 (a) Select Flat Surface (Pin), (b) Select Flat Surface (Ring)

After setup constraints, the assembly will be automatically orientied and the part becomes orange, which identifies Connection Definition Complete (Figure 7.24).

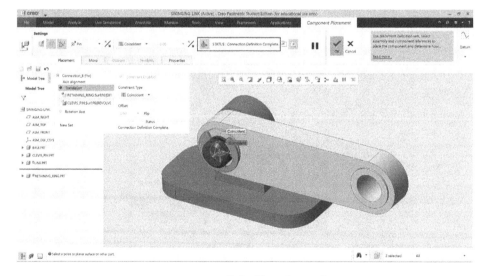

FIGURE 7.24 Coincident Constraint

Select OK to complete and return to view mode (Figure 7.25).

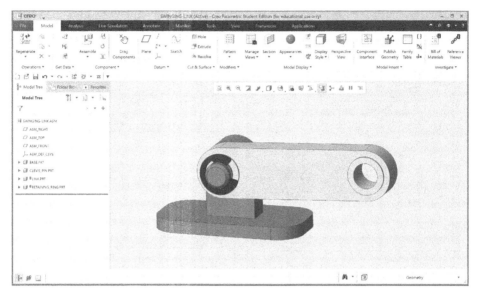

FIGURE 7.25   Review Assembly

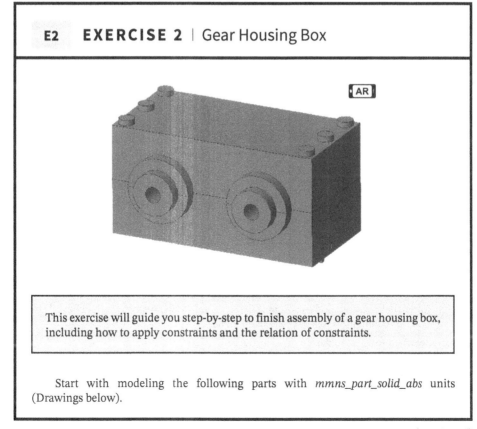

E2   EXERCISE 2 | Gear Housing Box

> This exercise will guide you step-by-step to finish assembly of a gear housing box, including how to apply constraints and the relation of constraints.

Start with modeling the following parts with *mmns_part_solid_abs* units (Drawings below).

(*continued*)

EXERCISE 2 | Gear Housing Box (*continued*)

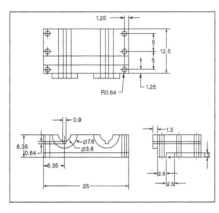
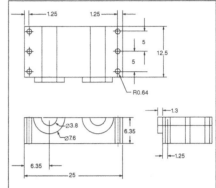
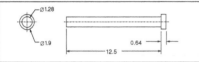
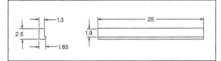

Individual Five Part Drawings (the 5th one is on the next page)

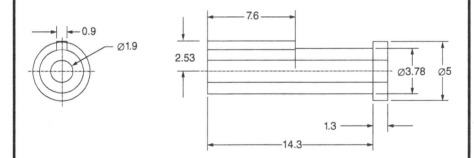

- Create New Project (or *Ctrl + N*).

  As the dialog box appears, ensure that "Assembly" type is selected (Figure 7.26a).

- Enter the assembly name (you may call it "Gear-Box-Housing") → Uncheck "Use default template" → Select OK (Figure 7.26a).

- Select "mmns_asm_design_abs" as the unit (Figure 7.26b).

E2   EXERCISE 2 | Gear Housing Box (*continued*)

Note: You may select different unit system based on how the parts are modeled.

(a)   (b)

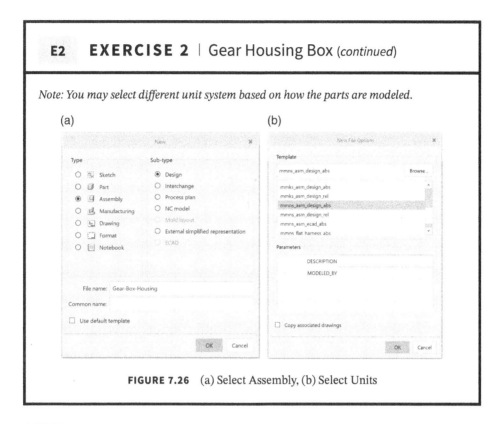

FIGURE 7.26   (a) Select Assembly, (b) Select Units

7.1.2 Start Assembly

The Assembly session will appear in another window (Figure 7.27) after creating the assembly.

- On the Model tab, select Assemble to add a component to the assembly (Figure 7.27).

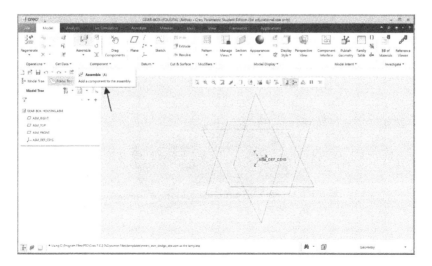

FIGURE 7.27   Assemble First Part

From here, a new dialog box "Open" (Figure 7.28) will appear. As we set the working directory to destination folder with completed parts earlier, here you may choose the first part to assemble (which is lower-housing.prt in this exercise).

FIGURE 7.28   Choose Part to Insert

- Select "lower-housing.prt" → Select OK.

Once you open the part, the Creo window will automatically switch to Component Placement tab (Figure 7.29). As the part is opened, you can use the middle mouse to rotate the part or Shift + Middle Mouse to move the part. Part orientation depends on how you modeled the parts previously.

- In the drop-down Constraints menu (Figure 7.29), choose Default.

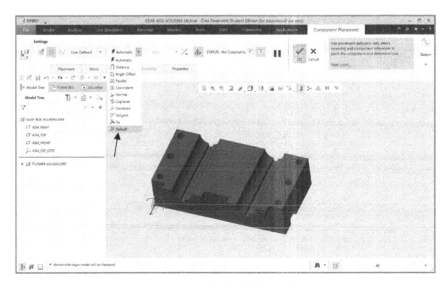

FIGURE 7.29   Default Constraints

After selecting Default, the part will turn orange color (Figure 7.30).
Note: Orange color identifies the part is Fully Defined.

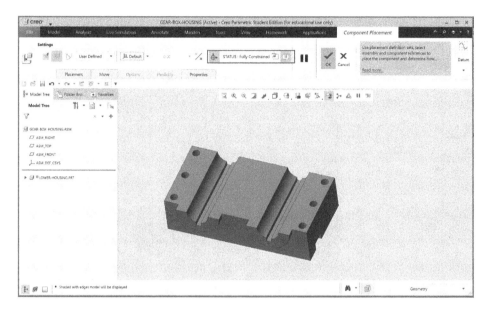

FIGURE 7.30   Verify Constraints

- Deselect showing axes and planes for clear view.
- Click OK to finish placing first part.

  Repeat the previous step to bring the second part into the assembly.

- Select Assemble to add next component (Figure 7.31).

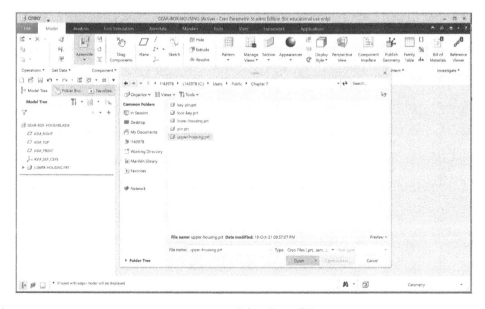

FIGURE 7.31   Select Second Part

- Select the upper housing (upper-housing.prt) → Select Open.

  Open Placement tab (Figure 7.32) to manage constraint details.

- Select first constraint as "Coincident" → Select 2 surfaces as shown in Figure 7.32.

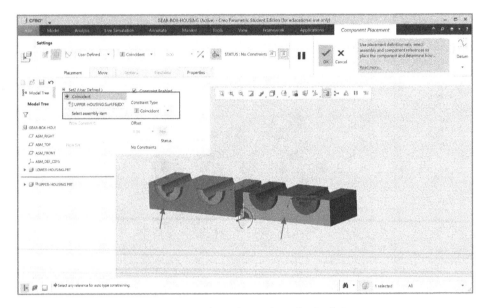

FIGURE 7.32 Select Flat Surfaces for Both Parts

- Select second constraint as "Coincident" → Select 2 surfaces as shown in Figure 7.33.

  Note: If the part is not automatically oriented as desired, select Change orientation of constraint (Figure 7.34) to flip the orientation.

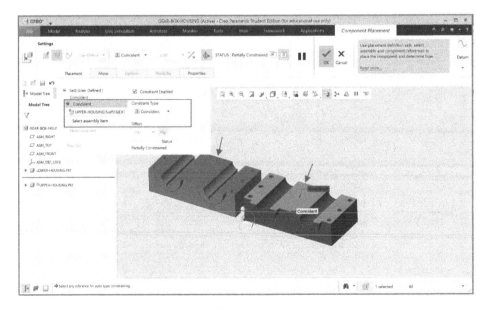

FIGURE 7.33 Flat Coincident Constraints

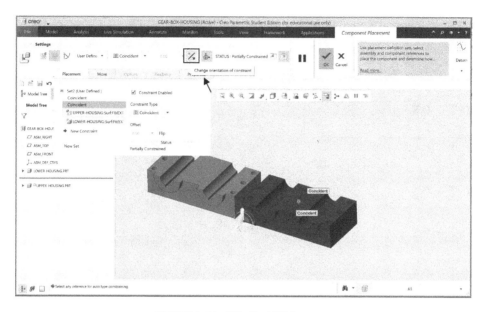

FIGURE 7.34   Flip Part If Necessary

The part will be automatically oriented as shown in Figure 7.35.

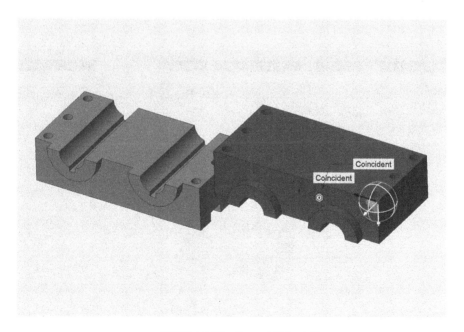

FIGURE 7.35   Correct Orientation

- Select the third constraint as "Coincident" → Select 2 surfaces as shown in Figure 7.36.

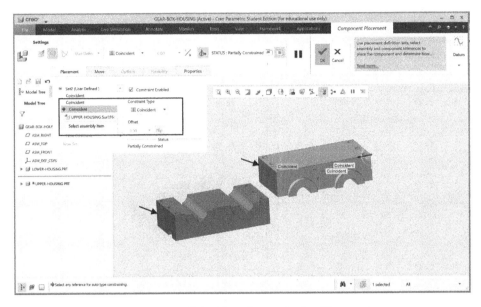

FIGURE 7.36   Coincident Constraint for Alignment

After setup constraints, the assembly will be automatically orientied and the part becomes orange, which identifies Connection Definition Complete (Figure 7.37).

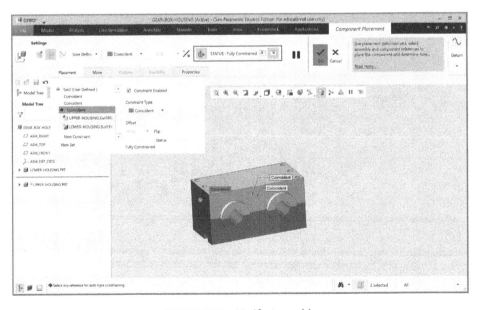

FIGURE 7.37   Verify Assembly

Select OK to complete and return to view mode (Figure 7.38).

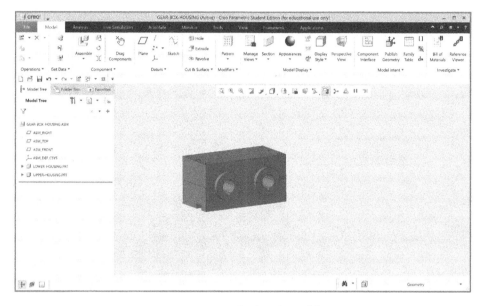

FIGURE 7.38   Review Assembly

7.1.3   Pattern Assembling

Repeat the previous step to bring the third part into the assembly.

- Select Assemble to add next component (Figure 7.39).
- Select Pin (pin.prt) → Select Open.

FIGURE 7.39   Insert Pin

Open Placement tab (Figure 7.40) to manage constraints easier.

- Select Translation (Figure 7.40) with "Coincident" as constraint type → Select 2 surfaces as shown in Figure 7.40.

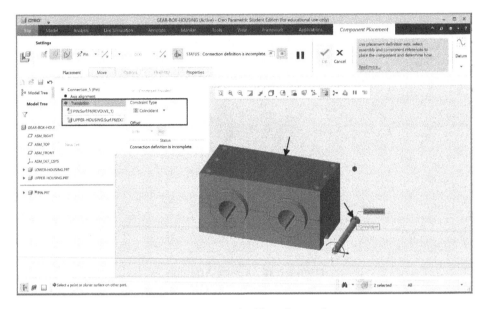

FIGURE 7.40 Coincident Constraint

- Select Axis alignment (Figure 7.41) with "Coincident" as constraint type.

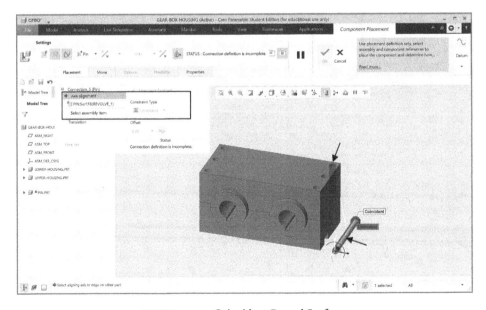

FIGURE 7.41 Coincident Round Surfaces

Select 2 surfaces as shown in Figure 7.41 → The part will be automatically oriented as shown in Figure 7.42.

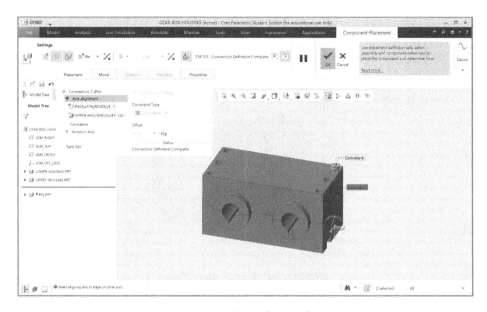

FIGURE 7.42   Correctly Mated Parts

Select the Pin on Model Tree (Figure 7.43), then select Patterns on Model tab (Figure 7.43).

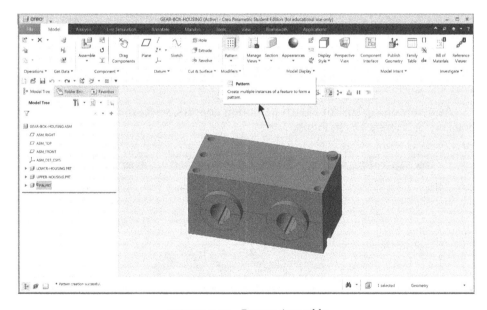

FIGURE 7.43   Pattern Assembly

Once Pattern tab is activated:

- In Select Pattern Type drop-down menu, select "Point" as Pattern Type (Figure 7.44).

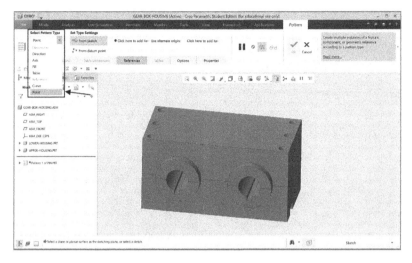

FIGURE 7.44 Point Pattern Type

- Click "Reference" to define the sketch for patterns (Figure 7.45).

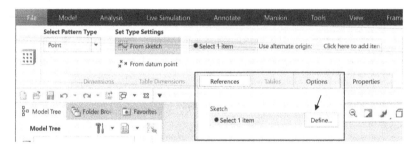

FIGURE 7.45 Define Reference

As a new dialog box for Sketch definition appears, select the top surface of the housing for reference as in Figure 7.46.

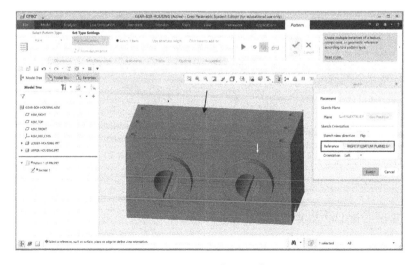

FIGURE 7.46 Select Surface

Once Sketch tab is activated (Figure 7.47), select Point at Datum (Figure 7.47) to sketch points for the patterns as shown in Figure 7.48.

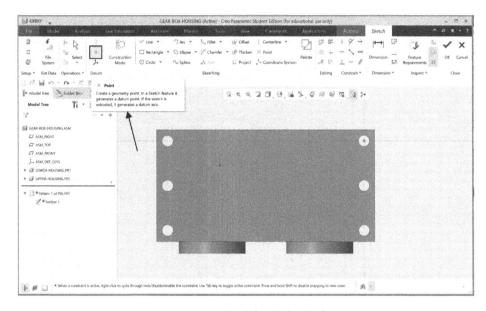

FIGURE 7.47   Select Point Tool

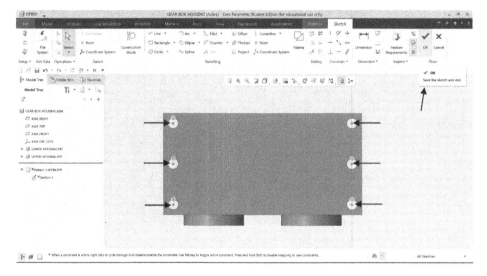

FIGURE 7.48   Define 6 Points

Sketch 6 points as in Figure 7.48 and then click OK to finish.

Once Sketch for pattern is defined, at the pattern tab, points of patterns will be shown as in Figure 7.49.

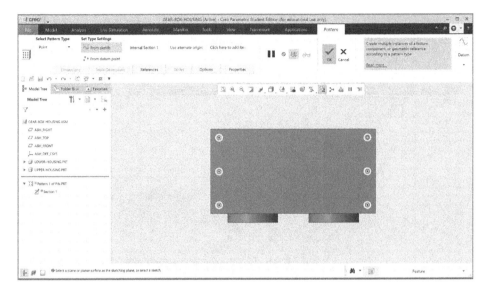

FIGURE 7.49 Verify Accuracy

Click OK to get back to Model view mode (Figure 7.50).

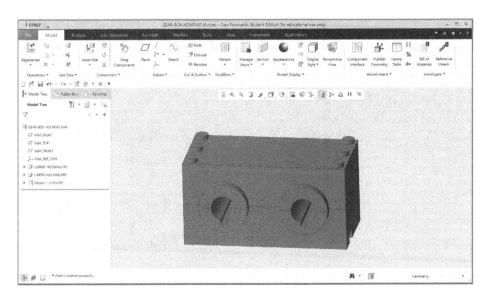

FIGURE 7.50 Review Assembly

Bring the next part into the assembly.

- Select Assemble to add next component (Figure 7.51).
- Select Key pin (key-pin.prt) → Select Open.

FIGURE 7.51   Insert Part

Open Placement tab (Figure 7.52) to manage constraint details.

- Select first constraint as "Coincident" → Select 2 surfaces as shown in Figure 7.52.

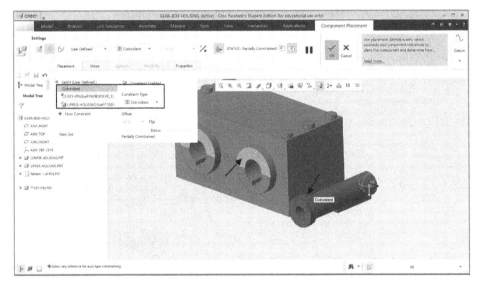

FIGURE 7.52   Flat Coincident Constraint

- Select second constraint as "Coincident" → Select 2 surfaces as shown in Figure 7.53.

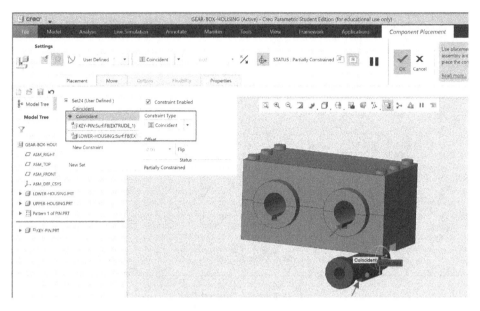

FIGURE 7.53 Select Mating Surface

- Select third constraint as "Coincident" → Select 2 surfaces as shown in Figure 7.54.

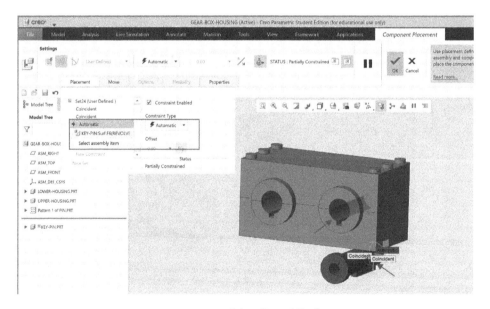

FIGURE 7.54 Select Round Surface

After required constraints are applied, the assembly will be as shown in Figure 7.55.

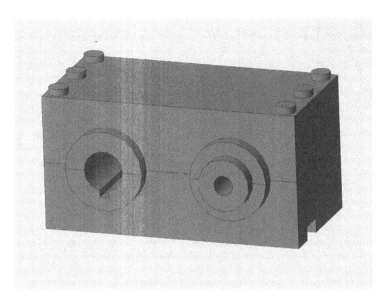

FIGURE 7.55   Review Assembly

Repeat previous steps to assemble the second key pin to get a final result as in Figure 7.56.

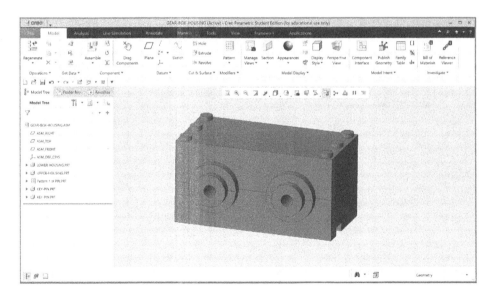

FIGURE 7.56   Review Completed Assembly

Bring the last part into the assembly.

- Select Assemble to add next component (Figure 7.57).
- Select Lock key (lock-key.prt) → Select Open.

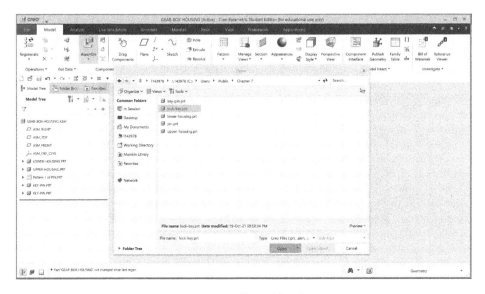

FIGURE 7.57 Insert New Part

Open Placement tab (Figure 7.58) to manage constraint details.

- Select first constraint as "Coincident" → Select 2 surfaces as shown in Figure 7.58.

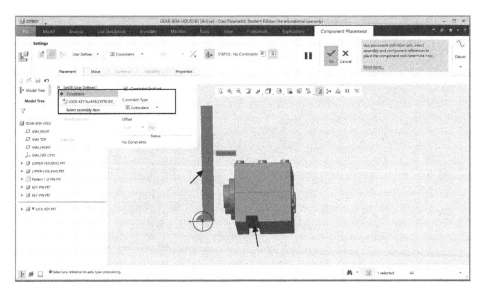

FIGURE 7.58 Select Flat Surface

- Select second constraint as "Coincident" → Select 2 surfaces as shown in Figure 7.59.

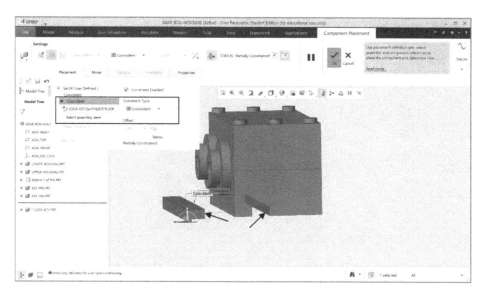

FIGURE 7.59   Select Mating Surface

- Select second constraint as "Coincident" → Select 2 surfaces as shown in Figure 7.60.

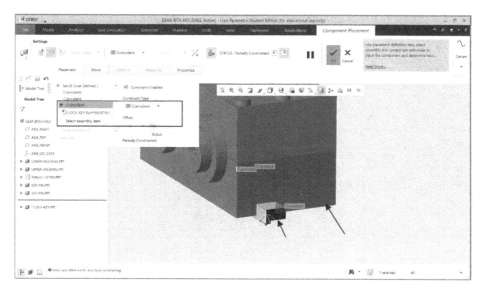

FIGURE 7.60   Coincident Constraint

- After required constraints are applied, the status of the lock key will become fully defined, and the assembly will be as shown in Figure 7.61.

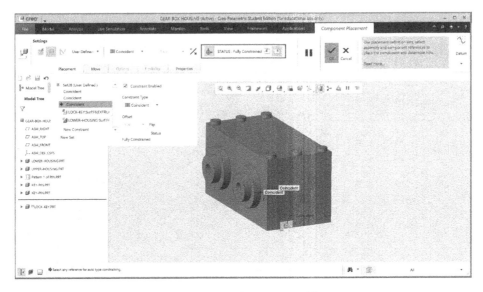

FIGURE 7.61 Accept Assembly

The final completed result will be shown as in Figure 7.62.

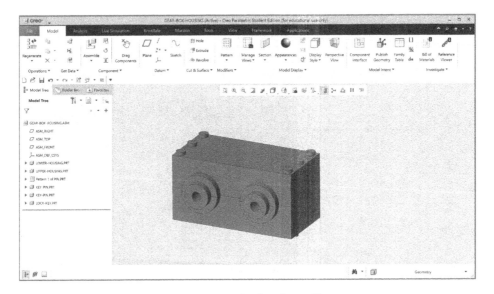

FIGURE 7.62 Review Assembly

Chapter Problems

This chapter problem will have number of sub assemblies to complete the master assembly for a prototype scooter by a start up company (Figure 7.63).

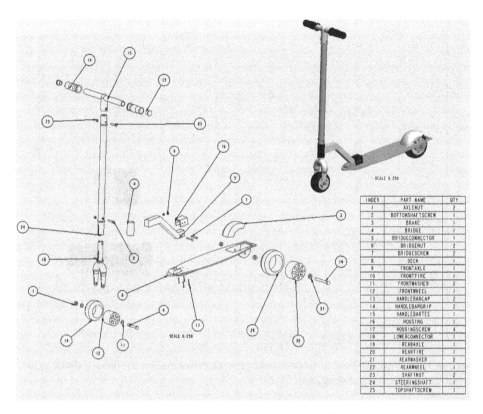

INDEX	PART NAME	QTY
1	AXLENUT	2
2	BOTTOMSHAFTSCREW	1
3	BRAKE	1
4	BRIDGE	1
5	BRIDGECONNECTOR	1
6	BRIDGENUT	2
7	BRIDGESCREW	2
8	DECK	1
9	FRONTAXLE	1
10	FRONTTIRE	1
11	FRONTWASHER	2
12	FRONTWHEEL	1
13	HANDLEBARCAP	2
14	HANDLEBARGRIP	2
15	HANDLEBARTEE	1
16	HOUSING	1
17	HOUSINGSCREW	4
18	LOWERCONNECTOR	1
19	REARAXLE	1
20	REARTIRE	1
21	REARWASHER	2
22	REARWHEEL	1
23	SHAFTNUT	2
24	STEERINGSHAFT	1
25	TOPSHAFTSCREW	1

FIGURE 7.63   Scooter Master Assembly

P7.1 Front Drive and Controls Assembly and Parts (Figures 7.64 and 7.65).

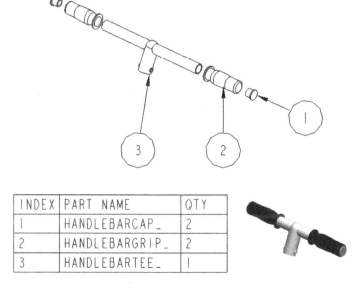

INDEX	PART NAME	QTY
1	HANDLEBARCAP_	2
2	HANDLEBARGRIP_	2
3	HANDLEBARTEE_	1

FIGURE 7.64   Front Drive Controls Assembly

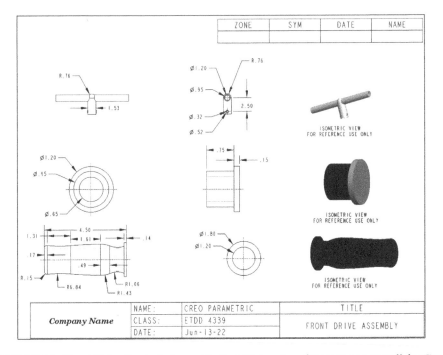

FIGURE 7.65   Front Drive Controls Assembly Part Dimensions (From Top: 1-HandlebarTee, 2-HandlebarCap, 3-HandlebarGrip)

P7.2 Model and Assemble STEM Assembly Components (Figures 7.66–7.68).

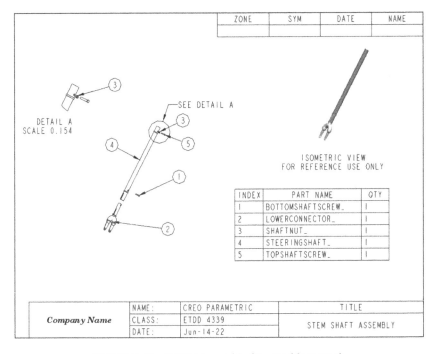

FIGURE 7.66   STEM Assembly (For Problem P7.2)

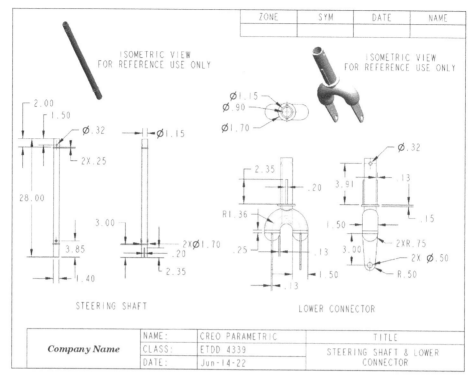

FIGURE 7.67 Steering Shaft and Lower Connector Dimensions (For Problem P7.2)

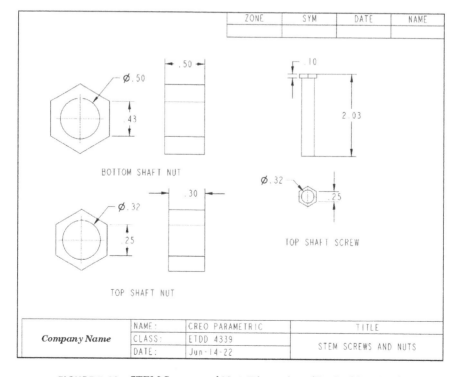

FIGURE 7.68 STEM Screws and Nuts Dimensions (For Problem P7.2)

P7.3 Model and Assemble Front Axle and Components (Axle Nut # 1 is same as Bottom Shaft Nut #2) (Figures 7.69 and 7.70).

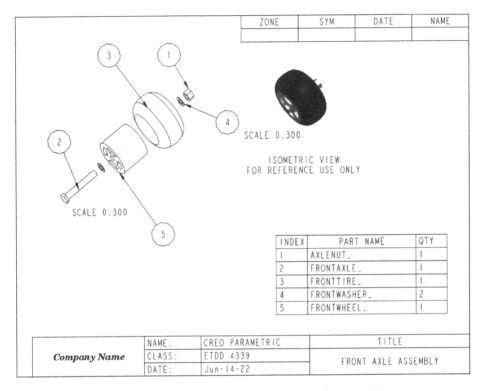

FIGURE 7.69 Front Axle Assembly (For Problem P7.3)

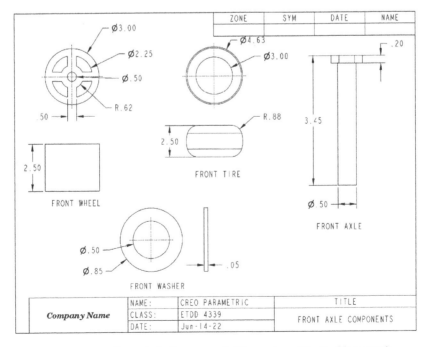

FIGURE 7.70 Front Axle Components Dimensions (For Problem P7.3)

P7.4 Model and Assemble Deck Components (Figures 7.71–7.74).

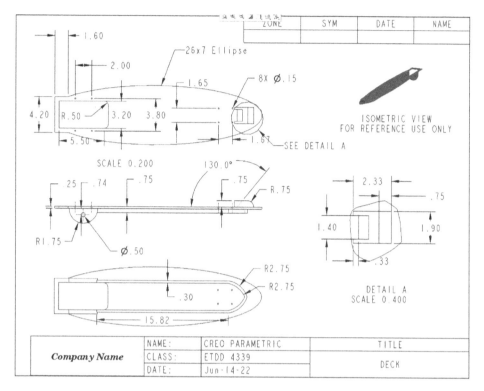

FIGURE 7.71 Deck Dimensions

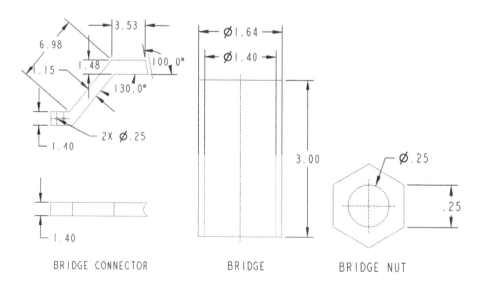

FIGURE 7.72 Deck Components Dimensions (For Problem P7.4)

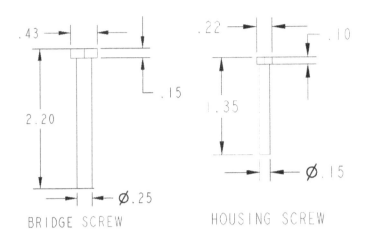

FIGURE 7.73 Deck Components Parts (For Problem P7.4)

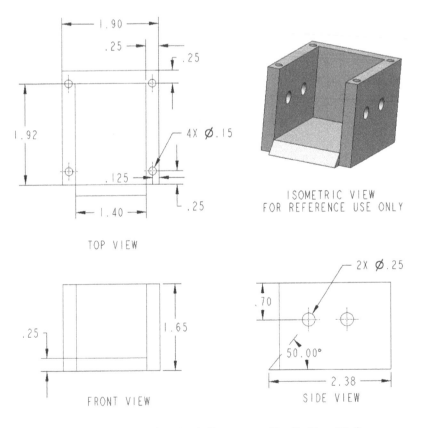

FIGURE 7.74 Housing, Deck Component (For Problem P7.4)

P7.5 Model and Assemble Scooter Rear Axle Components (Figures 7.75 and 7.76).
Note: Rear Axle Nut is same as the Front Axle Nut. Additionally, reuse Front Washer.

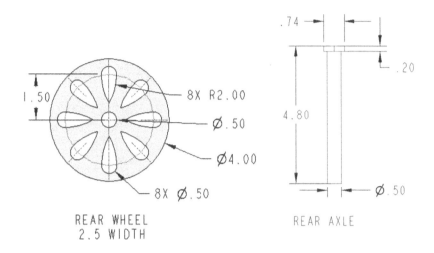

REAR WHEEL
2.5 WIDTH

REAR AXLE

FIGURE 7.75 Rear Axle Components (For Problem P7.5)

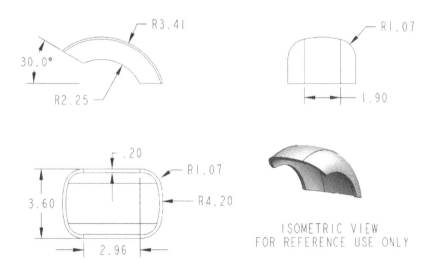

ISOMETRIC VIEW
FOR REFERENCE USE ONLY

FIGURE 7.76 Rear Axle Component-Brake (For Problem P7.5)

CHAPTER 8

Assembly Drawings

8.1 Introduction to ASM Drawings

Assembly drawings are used to represent components within the assembly in an exploded view. They show how individual components fit together, quantity of individual components within the assembly. Additionally, indexed annotations may also contain orthogonal views with sections. It is important to note that the assembly drawings should not duplicate the information that has already been reported such as component's individual drawing. You may hear assembly drawings referred to as overall assembly drawing, outline assembly drawing, exterior shape, diagrammatic assembly drawings, etc., but they all mean the assembly drawing with its individual components, broken down to quantities, indexes, and other necessary informational material.

CHAPTER OBJECTIVES

- Adjusting views of assembly
- The positions of components
- Manage views of assembly
- Adjust orientation of assembly
- Table of components of assembly
- Balloons of bill of materials
- Table of report of assembly
- Assembly drawing with designated format

Creo Parametric Modeling with Augmented Reality, First Edition. Ulan Dakeev.
© 2023 John Wiley & Sons, Inc. Published 2023 by John Wiley & Sons, Inc.

E1 **EXERCISE 1** | Swinging Link Assembly Drawing

This exercise will guide you step-by-step to finish the drawing of the swinging link assembly, including how to set up the orientation and component positions of the exploded view.

- Select working directory.

  Note: The working directory must contain all the parts used in the assembly and the assembly itself.

- Create New Project (or Ctrl + N).

  Note: As the dialog box appears, ensure that "Drawing" type is selected (Figure 8.1).

- Enter the assembly name (you may call it "Swinging-Link") → Uncheck "Use default template" → Select OK (Figure 8.1).

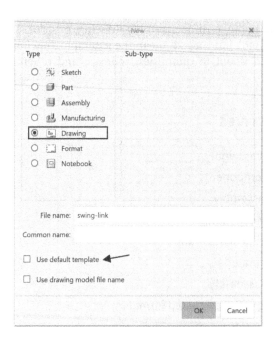

FIGURE 8.1 Select Drawing

- Select "Empty with format" in the Specify Template (Figure 8.2a) → Click "Browse. . .".
- Navigate to the folder containing the designated drawing template (drawingformat.frm for this exercise) → Select the template → Click Open (Figure 8.2b).

E1   **EXERCISE 1** | Swinging Link Assembly Drawing *(continued)*

Note: You may select different drawing template based on project requirements.

(a)   (b)

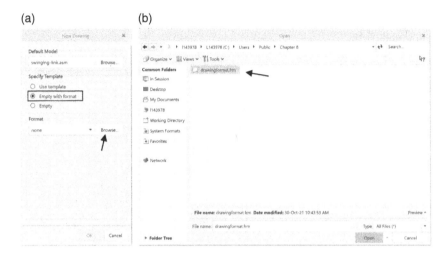

FIGURE 8.2   (a) Select Browse, (b) Select Drawing Format

As you select the designated template, the dialog box will show as Figure 8.3 → Select OK.

FIGURE 8.3   Verify Drawing Format Is Selected

(continued)

The Drawing session will appear in another window (Figure 8.4) after creating the drawing.

- On the Layout tab, select General View to add an assembly to the drawing (Figure 8.4).

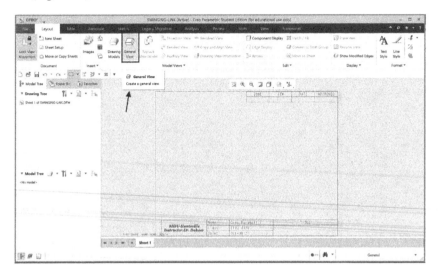

FIGURE 8.4 Insert General View

From here, a new dialog box "Open" (Figure 8.5) will appear. As we set the working directory to destination folder with completed parts earlier, here you may choose assembly to create drawing (which is swinging-link.asm in this exercise).

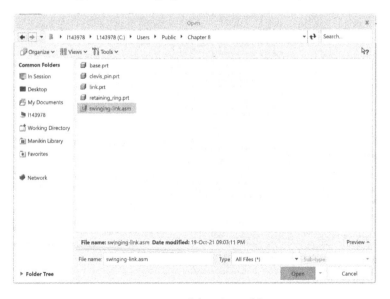

FIGURE 8.5 Select Assembly

- Select "swinging-link.asm" → Select OK.

A new dialog box "Select Combined State" will appear (Figure 8.6). Select Default All → Select OK.

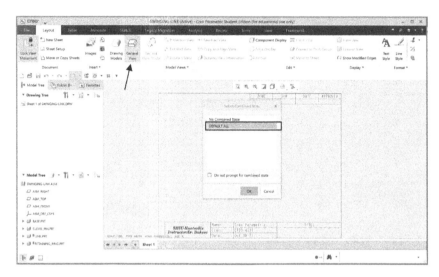

FIGURE 8.6   Default All

A new dialog box (Drawing View) will appear together with the assembly. As we have not set up orientation and exploded view for this assembly, the assembly view in the drawing will be shown incorrectly (Figure 8.7).

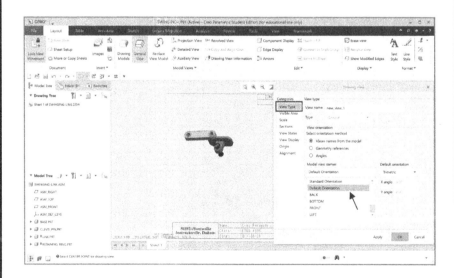

FIGURE 8.7   Select Orientation

(*continued*)

E1 **EXERCISE 1** | Swinging Link Assembly Drawing (*continued*)

To set up the assembly for the drawing, in the Model Tree, right-click the "SWINGING-LINK.ASM" → Select Open (Figure 8.8a) → As a new dialog box appears, select Default → Select Open (Figure 8.8b).

(a)   (b)

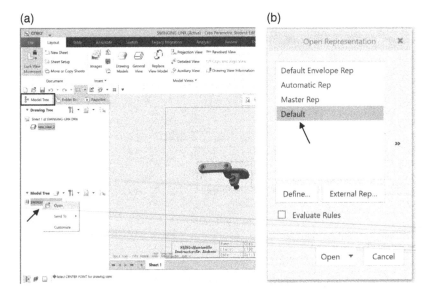

FIGURE 8.8   (a) Open Assembly, (b) Open Default

Creo will open an assembly session where we can set up orientation and exploded view (Figure 8.9).

- 🔲 Select "Edit Position".

Note: Make sure the Exploded View is selected (Figure 8.9).

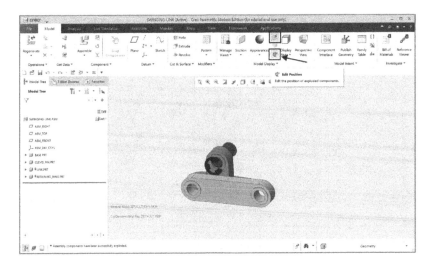

FIGURE 8.9   Edit Positions

E1   EXERCISE 1 | Swinging Link Assembly Drawing *(continued)*

Creo window will automatically switch to Explode Tool tab (Figure 8.10).

- Click at each component to activate the moving coordinator (Figure 8.10).

  Note: Make sure the moving mode is on.

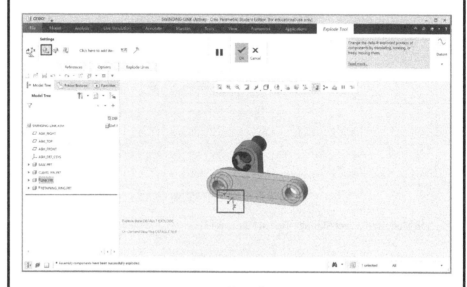

FIGURE 8.10   Move Components

- At this point, you can move the component by selecting one of the axes (it will turn green) and drag to the position where you want the component to be (Figure 8.11).

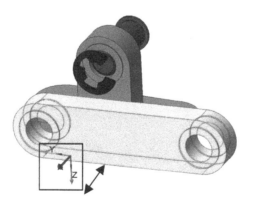

FIGURE 8.11   Spread Components Apart from Each Other

For the drawing to be tidy and clean, components cannot be overlapped with each other. Your final exploded view should look like Figure 8.12 → Select OK to save and close the Explode Tool tab.

(continued)

E1 **EXERCISE 1** | Swinging Link Assembly Drawing (*continued*)

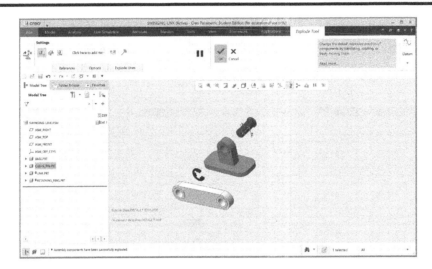

FIGURE 8.12 Verify Space

On Model tab, select Manage Views (Figure 8.13) to open the View Manager dialog box.

FIGURE 8.13 View Manager

When the View Manager is opened, make sure the explode tab is selected (Figure 8.14).

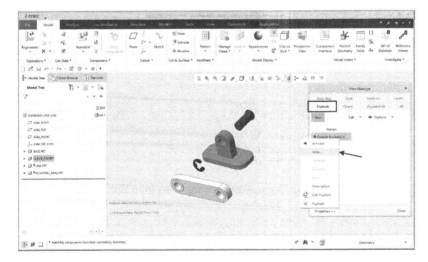

FIGURE 8.14 Select Explode Tab

E1 **EXERCISE 1** | Swinging Link Assembly Drawing *(continued)*

- Right-click at "Default Exploded" → Select "Save" to save the current exploded view.
- As the "Save Display Elements" dialog box appears → Click OK (Figure 8.15).

FIGURE 8.15 Save Default Explode

To set up orientation, in View Manager, switch to "Orient" tab (Figure 8.16a).

- Select New → Input the name (Isometric) of the new orientation (Figure 8.16b) → Press Enter → Click Close.

(a)   (b)

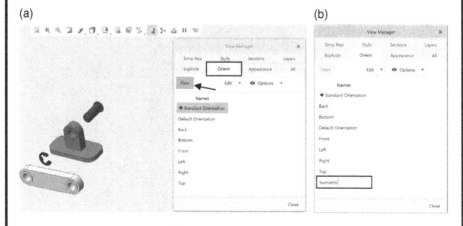

FIGURE 8.16 (a) Select Orient Tab, (b) Save Orientation

After finishing the setup, switch back to the drawing window:

- Select Windows button → Select Swinging-link.drw (Figure 8.17).

(continued)

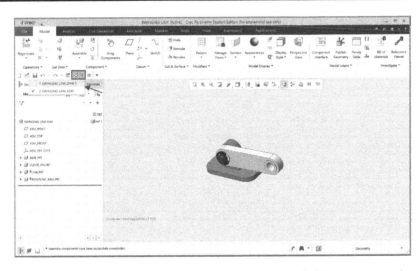

FIGURE 8.17 Change Windows (Note, Creo 9 Has Moved This Button to Top)

After switching back to the drawing window, double-click the assembly to activate Drawing View dialog box (Figure 8.18).

FIGURE 8.18 Select View Properties

In Drawing view Dialog box, set up as below:

- In "View Type" category, double-click the Isometric as Model view (Figure 8.19a).
- In "Scale" category, set Custom scale as 1.000 (Figure 8.19b).
- In "View Display" category, select No hidden as Display style (Figure 8.19c).
- Select OK to save and close the dialog box.

E1   EXERCISE 1 | Swinging Link Assembly Drawing (*continued*)

(a)

(b)

(c)

FIGURE 8.19   (a) Select Saved Orientation, (b) Scale View, (c) No Hidden

The final drawing with the exploded model will be as Figure 8.20.

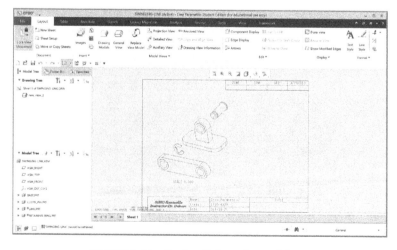

FIGURE 8.20   Review Orientation and State

8.1.1 Table of Report

In table tab, select Table → Create a 3×2 table (Figure 8.21) → Place the table in the drawing (Figure 8.22).

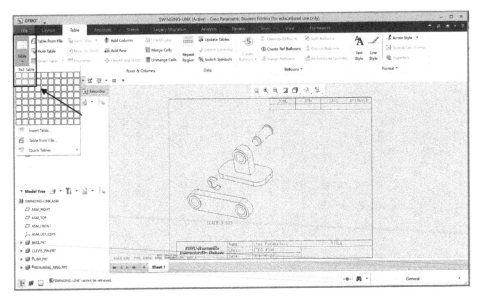

FIGURE 8.21 Insert Table

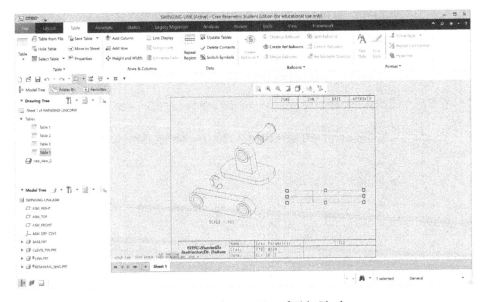

FIGURE 8.22 Place on Top of Title Block

Double-click in each cell on the first row to change name as shown in Figure 8.23.

INDEX	PART NAME	QTY

Change the width of the cell:

- Left-click the middle cell (PART NAME) → Select "Height and Width" (Figure 8.24a).
- In Height and Width dialog box, change Width to "10.000" → Select OK (Figure 8.24b).

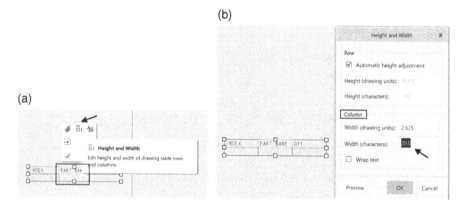

FIGURE 8.24   (a) Select Cell Size, (b) Adjust Cell Size

In the next steps, we add the repeat region to the table.

- Select the first cell of second row (Figure 8.25).
- Click the drop-down "Select Table" menu → Click "Select Row" (Figure 8.25).

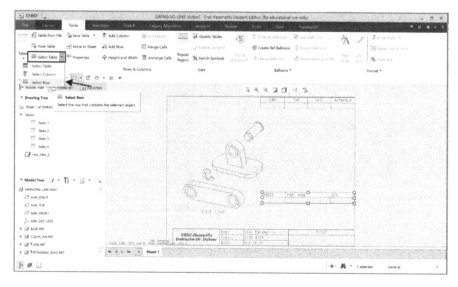

FIGURE 8.25   Select Row

- Right-click and hold the first cell of the second row (make sure it appears orange border as in Figure 8.26) to activate the properties menu (Figure 8.27).
- Select Add repeat region (Figure 8.27).

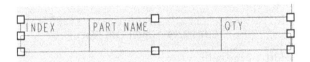

FIGURE 8.26 Right Click and Hold

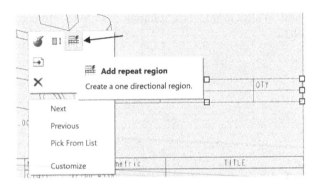

FIGURE 8.27 Add Repeat Region

Add the formula for the column, double-click the cell to activate the Report Symbol dialog box.

- For INDEX: Select rpt. . . (Figure 8.28a) → Select index (Figure 8.28b).

FIGURE 8.28 (a) First Cell Report, (b) Select Index

- For QTY: Select rpt... (Figure 8.29a) → Select qty (Figure 8.29b).

FIGURE 8.29   (a) Third Cell Report, (b) Report Quantity

- For PART NAME: Select asm... (Figure 8.30a) → Select mbr... (Figure 8.30b) → Select name... (Figure 8.30c).

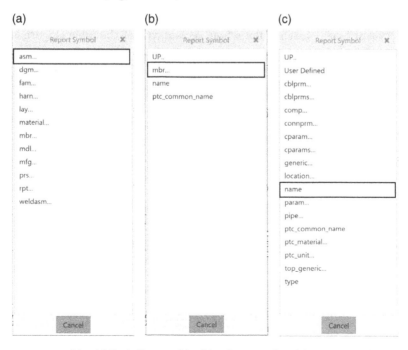

FIGURE 8.30   (a) Middle Cell Assembly, (b) Select Member, (c) Select Member Name

- The final table should look like Figure 8.31.

INDEX	PART NAME	QTY
rpt.index	asm.mbr.name	rpt.qty

FIGURE 8.31 Verify Cells

Click Repeat Region in Data menu (Figure 8.32) to activate the Menu Manager.

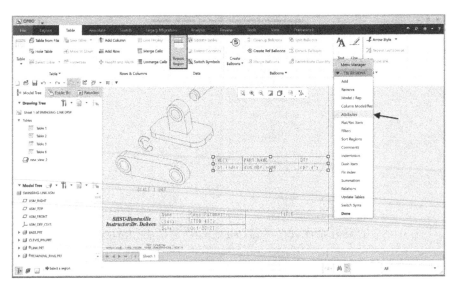

FIGURE 8.32 Select Repeat Region and Attributes

Select Attributes → Select the whole second row of the table (Figure 8.33) → Click Done.

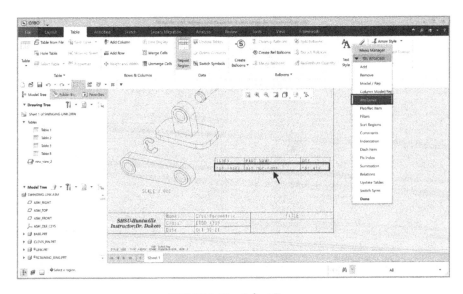

FIGURE 8.33 Select Row

In Region Attribute tab, Select No Duplicates → Select Done/Return (Figure 8.34). The table now will show as Figure 8.34.

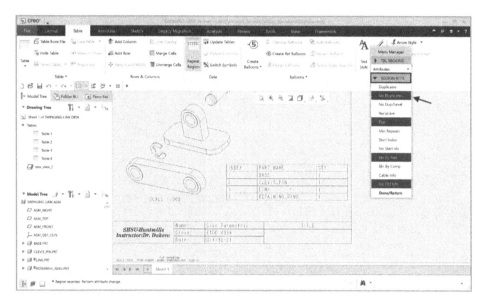

FIGURE 8.34   No Duplicates

To create Balloons, Click the Create Balloons button (Figure 8.35) → Select "Create Balloon – All".

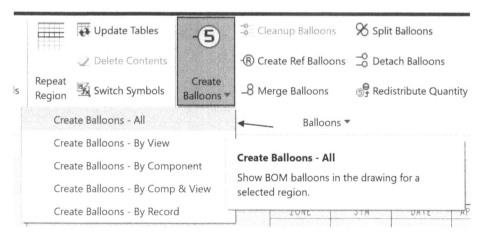

FIGURE 8.35   Select Create Balloons – All

As the balloons appear, make sure that the balloons are not overlapped with each other and with the components (Figure 8.36).

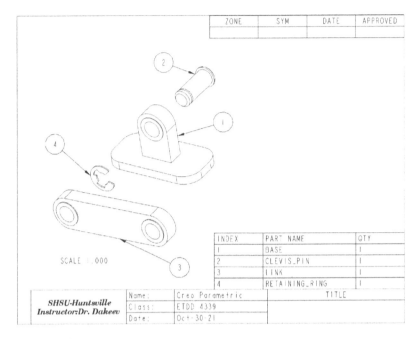

FIGURE 8.36   Review Balloons

8.1.2   Create Isometric view

Click at General View to add another isometric view to the drawing (Figure 8.37) →
Select Default All → Click OK.

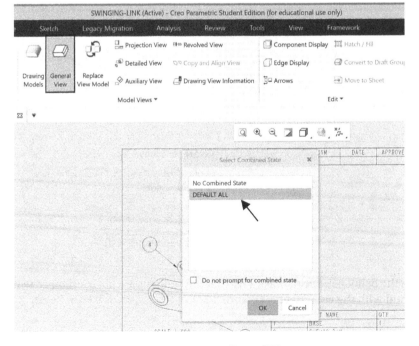

FIGURE 8.37   Insert General View

In Drawing view Dialog box, set up as below:

- In "View Type" category, double-click the Isometric as Model view (Figure 8.38a).
- In "Scale" category, set Custom scale as 1.000 (Figure 8.38b).
- In "View State" category, uncheck "Explode components in view" (Figure 8.38c).
- In "View Display" category, select Shading as Display style (Figure 8.38d).
- Select OK to save and close the dialog box.

(a)   (b)

(c)   (d)

FIGURE 8.38   (a) Select Isometric Orientation, (b) Scale as Needed, (c) Select View States, (d) Select Shading

To move the view, uncheck the "Lock View Movement" (Figure 8.39).

- Add the Title by double-clicking the cell → Input the title (*"SWINGING LINK"*).

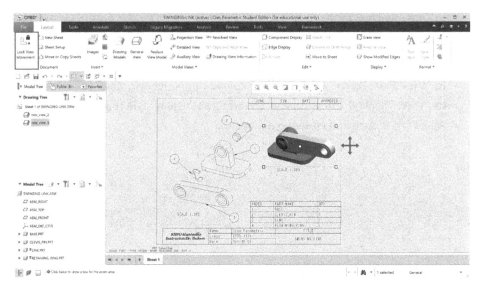

FIGURE 8.39 Unlock View

The final drawing should appear like Figure 8.40.

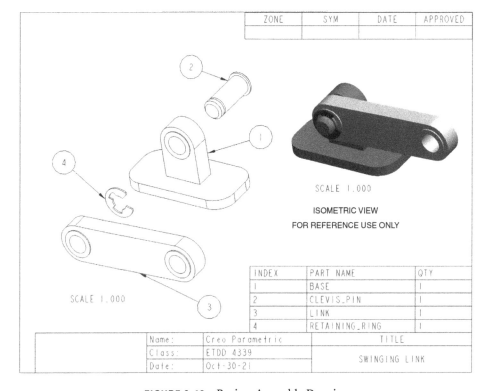

FIGURE 8.40 Review Assembly Drawing

Chapter Problems

P8.1 Develop a drawing for Front Drive & Controls Assembly on Problem 7.1 (Figure 8.41).

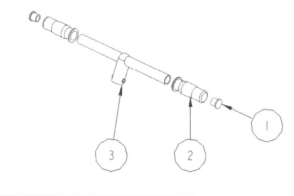

INDEX	PART NAME	QTY
1	HANDLEBARCAP_	2
2	HANDLEBARGRIP_	2
3	HANDLEBARTEE_	1

FIGURE 8.41   Front Drive Controls Assembly Drawing

P8.2 Develop a drawing for STEM Assembly Components on Problem 7.2 (Figure 8.42).

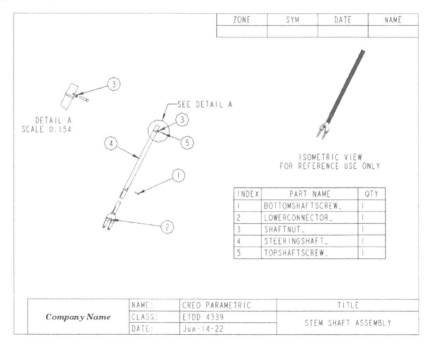

FIGURE 8.42   STEM Assembly Components Drawing

P8.3 Develop a drawing for Front Axle Components on Problem 7.3 (Figure 8.43).

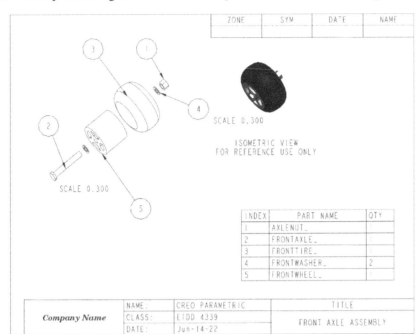

FIGURE 8.43 Front Axle Components Drawing

P8.4 Develop a drawing for Deck Components.

P8.5 Develop a master assembly for the Scooter project (*Note: you can assemble individual assemblies as if you are assembling an individual component*) (Figure 8.44).

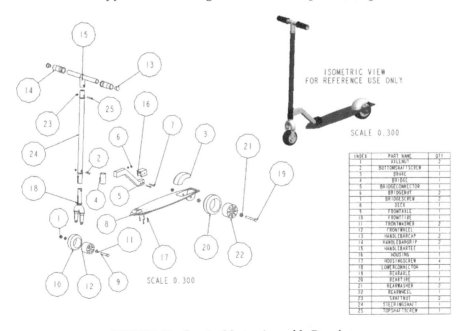

FIGURE 8.44 Scooter Master Assembly Drawing

CHAPTER 9

Geometric Dimensioning and Tolerancing GD&T

Introduction

A drawing is a communication tool for the part or assembly or any other system so that the stakeholders can understand each other clearly and accurately, which is the core driving force of industries. We communicate by indicating two elements in all drawings, which are 1-shape, and 2-size. Until now, we have been discussing size for parts through their dimensional values, annotations, scales etc. However, we never mentioned the shape of a part. Sure, logically we should be able to indicate how a part should look like from its isometric or other views, but when it comes to precision manufacturing with robotic arms, plasma cutters, or gauging tools, we must accurately declare how the part's shape should look like.

CHAPTER OBJECTIVES

After completing this chapter, you should:

- Understand tolerances.
- Understand Geometric Dimensions and Tolerances (GD&T)
- Insert Control Frame
- Understand GD&T Symbols
- Insert Datums and Tags
- Understand Material Conditions
- Learn how to Interpret GD&T Symbols

For instance, let us consider the following circular hole that was manufacturing with a drilling tool (Figure 9.1).

Creo Parametric Modeling with Augmented Reality, First Edition. Ulan Dakeev.
© 2023 John Wiley & Sons, Inc. Published 2023 by John Wiley & Sons, Inc.

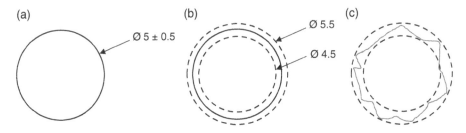

FIGURE 9.1 (a) Required Hole Diameter, (b) Acceptable Sizes, (c) Acceptable Sizes

From this dimensional value, we see that the customer wants a hole with a diameter of size between 4.5 and 5.5.

Therefore, the supplier can provide the parts within these dimensions. So, a drilled hole with 4.7 diameter is an acceptable part, right?

So, what if the supplier provides the following part? Does the customer accept it? First, answer to this question is automatic "NO". However, based on our acceptable values, this part fits the requirement and we, as a customer, must accept this part.

The problem is this part will not fit our final assembly because a mating part (consider a pin that will be inserted into this hole) with 4.8 diameter is almost impossible to be the same shape. Consider we ordered 2 million parts for 30 cents each? Our organization will be losing $600 000 because all of these will need to go to scrap. This is the reason; we need to communicate the shape of the part to the supplier. And we communicate the shape through geometric dimensions and tolerances (GD&T).

Therefore, GD&T is communication symbols that refer to shape, size, and positional relationships to the mating part or the part itself, and then tolerance is an allowed error since not all machinery can produce ideally. The machines may not hold the required tolerances because it might be out of calibration, or the equipment might be older model that is not capable of producing parts with tight tolerances.

There are two approaches to classify drawing design, 1-size tolerance, and 2-geometric tolerance (size and shape as mentioned earlier). Therefore, GD&T provides a clear and concise technique for defining reference coordinates (datums) and dramatically reduce the need for drawing notes. In this chapter, we will learn how to apply GD&T symbols on part drawings.

In this regard, GD&T is a-Symbols, b-Rules, c-Mathematical Definition (Y14.5.1), 4-Internationally recognized standard (ASME Y14.5.1 and ISO 1101).

Because tolerance refers to an error, or how much can a manufactured part's size and shape can deviate from the provided 3D model, there are three types of tolerances we need to discuss:

Dimensional tolerance refers how much a part's size can deviate from its ideal measures (a totally flat surface of a dining table may not be glass smooth but with micro-level dimples etc.)

Form tolerance is part of the shape (two S's, shape, and size), which refers how much the manufactured part can deviate from its ideal geometric shape (the tabletop should not be sloped to one side of the table).

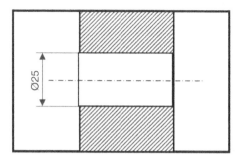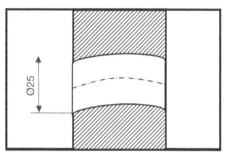

FIGURE 9.2 Required Geometric Dimension of the Hole vs Deviated Shape with the Accurate Diameter

Position tolerance ensures how much error is acceptable from the ideal location of a feature (a hole on the table should not be excessively positioned in a wrong location, Figure 9.2).

9.1.1 GD&T Symbols

Because GD&T is a symbolic language on engineering drawings and models that communicate how much deviation is allowed on a part's geometry, the production team can easily understand the design intent and specification requirements of the customer. If two persons with different languages try to communicate, there will be lots of errors in their discussion. Therefore, for any language to be effective, it must follow a common standard, which is American Society for Mechanical Engineering (ASME, Table 9.1) that is accepted and adopted in the US and ISO standards for European countries. These symbols are periodically revised by the subject matter experts to ensure the final product is manufactured accurately.

In this chapter exercises, we will be introducing GD&T feature control frames (Figure 9.3) that communicate what should be considered during the manufacturing of a part. The feature control frame can be attached directly to a feature or a dimension. We will be attaching to the dimension without a leader. This feature control frame is read as "this feature has a true position of 0.03 diameter at maximum material condition and is referenced to A, B, and C datums respectively".

Figure 9.3 also contains maximum a term we come across commonly, that is material condition symbol. Below are other often used symbols in GD&T (Figure 9.4).

Consider a plate A, where we drilled a 0.5 diameter hole and a second plate B of same size with 0.7 diameter hole. Which one of these plates will have maximum material in it? Surely, plate A will have maximum material because a smaller diameter hole removes less amount of material from the product. Similarly, plate B will have least material condition (LMC) because more material has been removed from the plate due to larger diameter hole. Therefore, the symbol MMC indicates that the tolerance is allowed only if the product is at maximum material condition or LMC.

TABLE 9.1 GD&T Symbols and Their Descriptions

Tolerance Type	Description	Symbol	ASME Section
Form	Straightness	——	6.4.1
	Flatness	▱	6.4.2
	Circularity	◯	6.4.3
Profile	Line profile	⌒	6.5.2(b)
	Surface profile	⌓	6.5.2(a)
Orientation	Angularity	∠	6.6.2
	Perpendicularity	⊥	6.6.4
	Parallelism	//	6.6.3
Location	Position	⊕	5.2
	Concentricity	◎	5.11.3
	Symmetry	⩵	5.13
Runout	Circular runout*	↗	6.7.1.2.1
	Total runout*	↗↗	6.7.1.2.2

*Circular runout controls only a particular circular cross section of a part, while total runout controls the entire surface of the part.

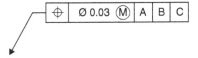

FIGURE 9.3 Control Frame with GD&T Symbols with a Leader

Definition	Symbol	ASME Section
At maximum material condition (MMC)	Ⓜ	3.3.5
At least material condition (LMC)	Ⓛ	3.3.5
Diameter	⌀	3.3.7
Between*	↔	3.3.11
Reference	( )	3.3.8
Radius	R	3.3.7
Square	□	3.3.15
Centerline	℄	
Places or by	8X	

FIGURE 9.4 Frequently Used GD&T Symbols

Table 9.2 shows additional visual demonstration of various tolerance deviations from the part's nominal design:

TABLE 9.2 **Types of Errors (Deviations) That May Occur During the Part Production**

Design Deviations	True Deviation
Geometrically ideal (required) nominal design	Nominal design
Dimensional deviation	Dimensional deviation
Shape deviation	Shape deviation
Positions deviations	Location deviation Orientation deviation
Surface	Surface deviation

E1 **EXERCISE 1** | Base Drawing (Size Tolerance)

Let us practice GD&T in practice with opening a previously developed drawing for the base part of swinging link. If you do not have the drawing, you can quickly develop one

(*continued*)

and save it in the working directory. Note that the part must also be placed into the same working directory with the drawing.

- Open (or develop) Base part Drawing for swinging link assembly (Figure 9.5).

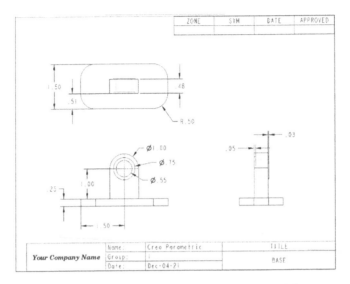

FIGURE 9.5   Base Part Drawing for Swinging Link

Previously, we annotated features with their nominal dimensional values. Because it is impossible to manufacture a perfectly accurate part continuously in the industry, we will need to provide acceptable tolerances. In Creo Parametric, we will need to make a quick configuration change on the drawing to introduce size tolerances.

- Click File → Prepare → Drawing Properties (Figure 9.6).

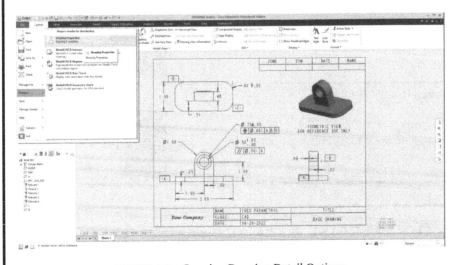

FIGURE 9.6   Opening Drawing Detail Options

E1   **EXERCISE 1** | Base Drawing (Size Tolerance) *(continued)*

- Click "change" for the Detail Options (Figure 9.7).

FIGURE 9.7   Modify Detail Options

- Type "tol_display" in the search box (Figure 9.8) ➜ Change value from "no" to "yes" ➜ Click Add/Change to accept ➜ Close Drawing Options.

FIGURE 9.8   Change Tolerance Display Value

9.1.2 Dimension Tolerance

As Tolerance is active, we can now activate the tolerance for each dimension.

Click at the dimension → Select Tolerance drop-down menu → Select Symmetric.

- Select "0.75" diameter dimension → On the Dimension Ribbon, click "Tolerance" drop down (notice that this button was not available before we changed the configuration) → Select "Symmetric" from the list (Figure 9.9).

We will indicate that the ring for the base part with the diameter 0.75 will be acceptable with ±0.05 value. That means any base part supplied to us within 0.745 and 0.755 diameter ring sizes will be used in the assembly because it does not affect the functionality of the final product if the size changes slightly. To show it on the drawing, we will need to enter the tolerance value as shown in Figure 9.10.

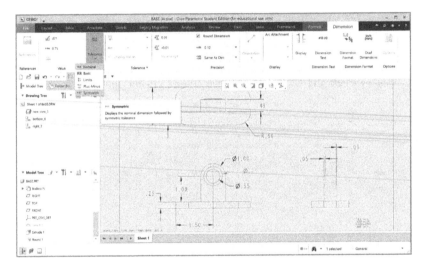

FIGURE 9.9 Applying "Symmetric" Tolerance Type

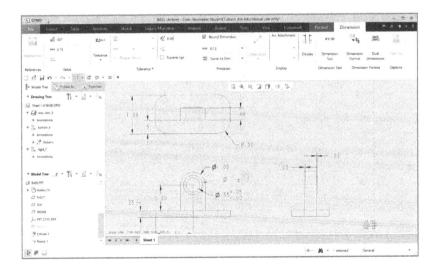

FIGURE 9.10 Defining 0.05 Symmetric Tolerance Value to the 0.75 Diameter Ring

Similarly, we will apply Plus-Minus tolerance to 0.55 diameter hole to indicate that we are OK accepting a slightly larger hole size (up to 0.05 diameter from the nominal value), but we will not accept any smaller hole.

- Select 0.55 diameter dimension → Click Tolerance tool → Select Plus-Minus from the list → enter 0.05 for plus tolerance and 0 for the minus value (Figures 9.11).

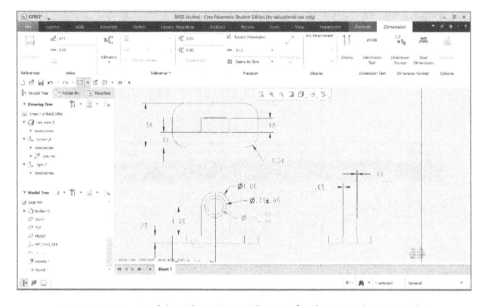

FIGURE 9.11   Applying Plus-Minus Tolerance for the 0.55 Diameter Hole

---

E2   EXERCISE 2 | Base Drawing (Geometric Tolerance)

- Under "Annotate" tab, select Geometric Tolerance tool (Figure 9.12a,b) → Select 0.75 diameter dimension value to attach the feature control frame.
- Enter the following GD&T values in the feature control frame (Figure 9.13).

So, what did we do? We indicated that the 0.75 diameter ring with 0.05 symmetrical tolerance is good for production as a size. We also added that the ring should be within 0.001 diameter positional tolerance from its true position (we would not want the correct size hole to shift sideways). But where is the true position? We indicated that the true position will be referenced first from datum A, then datum B, and datum C.

(continued)

E2 EXERCISE 2 | Base Drawing (Geometric Tolerance)

(continued)

(a)

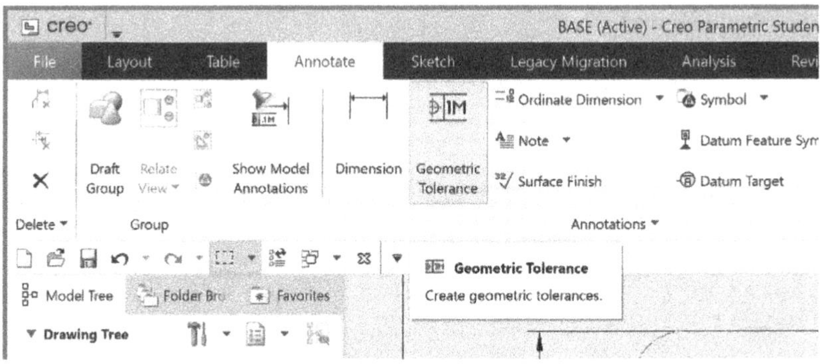

(b)

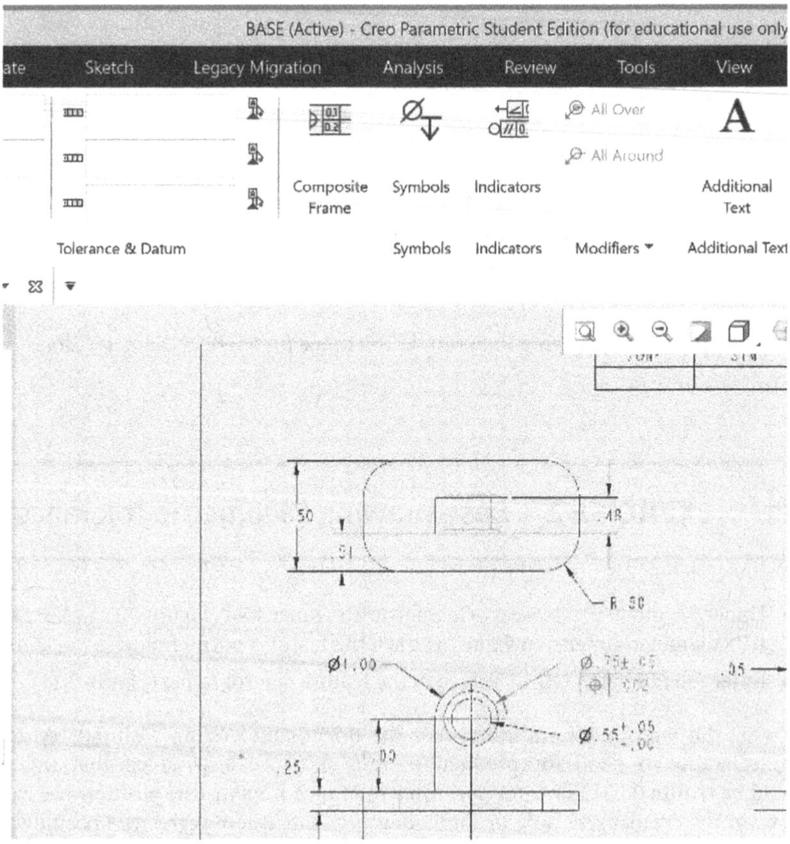

FIGURE 9.12 (a) Geometric Tolerance Tool on an Existing Dimension, (b) Geometric Tolerance Tool on an Existing Dimension

---

E2 **EXERCISE 2** | Base Drawing (Geometric Tolerance)

(*continued*)

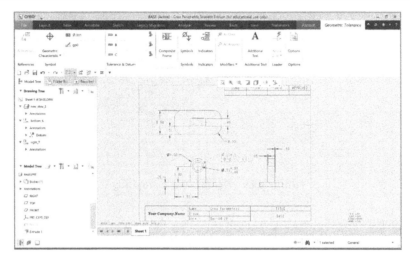

FIGURE 9.13   GD&T Symbols in the Feature Control Frame

9.1.3 Add Datum References

There are four datum plane references for the base part: datum plane A – bottom surface of the part, datum plane B – front surface of the base feature, datum plane C – the left side surface of the base, and datum axis D-the hole axis. To add these datum references, we need to switch to the part and indicate in the datum properties.

- Open the Base part ➔ Select the bottom plane (if it does not exist, you can always introduce a new plane by clicking "Plane" on the datum panel and selecting the surface you want to place it on) ➔ Right-click and hold and click on properties (Figure 9.14) ➔ Type "A" to rename ➔ Select A datum graphics inside the rectangle (Figure 9.15).

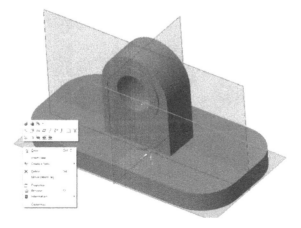

FIGURE 9.14   Datum Properties

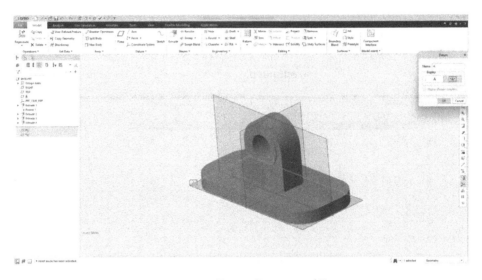

FIGURE 9.15 Datum Rename and Tag

Notice that the datum tag "A" appears on the datum plane (Figure 9.14).

- Click OK to accept the datum tag properties.
- Repeat this for datum planes B and C.
- To tag axis datum, right-click on the axis ➔ Rename axis datum ➔ Apply datum axis tag (same steps as adding datum planes and tags) as shown in Figure 9.16.

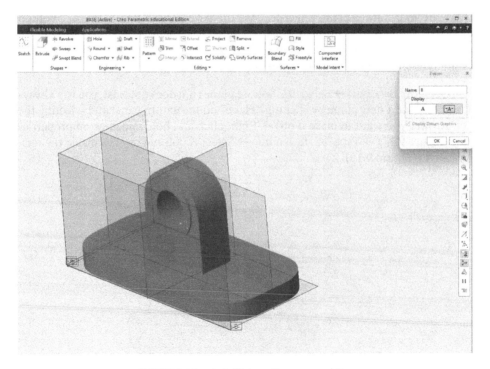

FIGURE 9.16 Axis Datum Rename and Tag

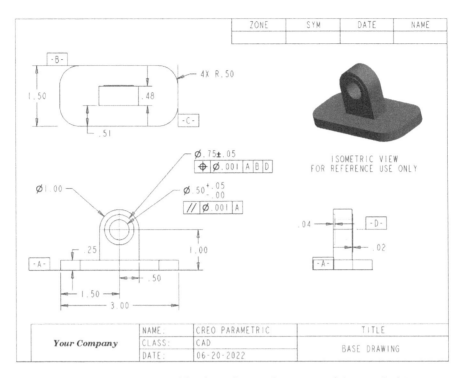

FIGURE 9.17 Base Part Drawing with Dimensions, Tolerances, and Geometrical Feature Control Frames

- Switch back to drawing ➔ Delete datums that do not belong to a view (Figure 9.17) ➔ Add the rest of the GD&T information to complete the drawing.

  Note: To modify datums on the drawing (move, delete, etc.), you need to be on the "Annotate" tab.

9.1.4 Knowledge Check Activity

1. Interpret GD&T notes attached to the 0.5 diameter hole.
2. Add a feature control frame to communicate the front face of the base. The surface must be flat within 0.01 allowable tolerance with respect to datum plane B.

E3   EXERCISE 3 | Additional Size and Form Tolerance Practices

A supplier company requested GD&T communication on a suspension arm bracket (Figure 9.18), which was provided to them in metric units. You, as design engineer, need to develop a drawing with your company's format that includes the following size and form tolerances: 1 – All 16 mm holes must have 0.125 symmetrical tolerances.

(continued)

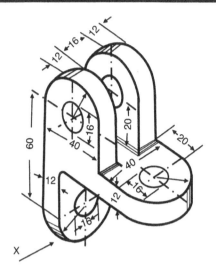

FIGURE 9.18 Bracket

Side holes must have Ø0.001 positional tolerance references from bottom surface as primary, and B datum plane as secondary at maximum material condition. Top View hole must reference from secondary datums B and tertiary datum C. Additionally, you need to tell the supplier company that the part's side face (datum plane C, X arrow) can vary between 0.001 flatness tolerance at least material condition.

- Start a new drawing → Insert Top, Front, and Side views along with the Isometric View.
- Activate Tolerance Display in the detail → Add 0.125 symmetric tolerance to all Ø16 holes.

  Note: If 0.125 is rounding to 0.13, uncheck "Round Dimension" (Figure 9.19) next to 0.125 value you entered.

- Under "Annotate" tab, click Geometric Tolerance → Click the Ø16 diameter hole (Top View) dimension with its tolerance values (Figure 9.19) to attach → Enter positional geometric characteristic with Ø0.001 at MMC and datum plane references (Figure 9.20).

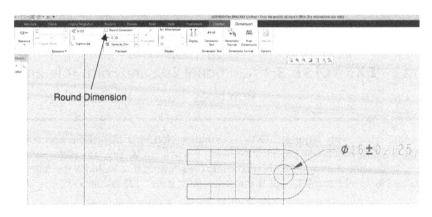

FIGURE 9.19 Axis Datum Rename and Tag

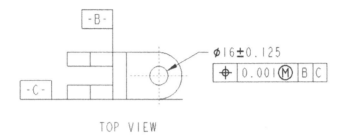

TOP VIEW

FIGURE 9.20   Feature Control Frame

- Repeat the step above to add a feature control frame with its values for the side view hole (Figure 9.21).

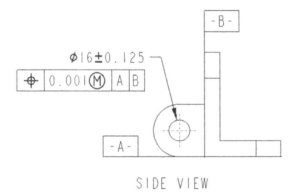

SIDE VIEW

FIGURE 9.21   Feature Control Frame for Side View

- Add flatness geometric characteristic for the front face of the part (Figure 9.22).

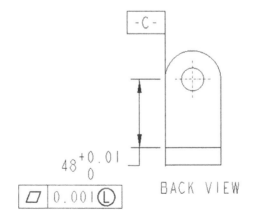

BACK VIEW

FIGURE 9.22   Feature Control Frame for Back View

Complete the rest of the drawing with required dimensions with nominal tolerance values

Chapter Problems

P9.1 Model a connection plate with the following size and shape tolerances. Consider the datum placements for accurate results (Figure 9.23).

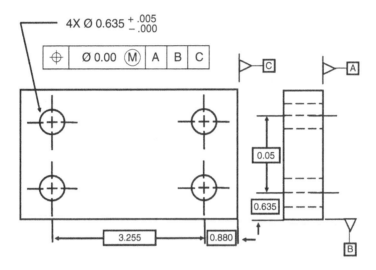

FIGURE 9.23 Assignment P9.1

P9.2 Model Stock housing 165 below. Add 0.05 symmetric tolerance at bottom surface, where 96 dimension is located and a flatness tolerance 0.01 on datum A (the bottom surface), 0.05 plus and 0.00 minus tolerance on datum B (64 dimension), and add a tertiary datum X on the front face with X arrow (Figure 9.24).

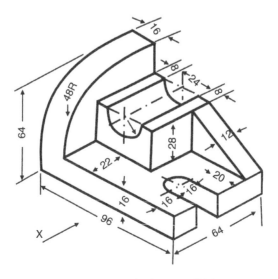

FIGURE 9.24 Assignment P9.2

P9.3 Develop mold housing 796 with both size and shape tolerances. Add 0.03 symmetric tolerance on 80 and 40 (datum X on the front surface with X arrow), 0.01 diameter circularity tolerance on top curve, and add Y datum at the bottom surface. All remaining dimensions must have 0.02 plus tolerance and 0.00 minus tolerance (Figure 9.25).

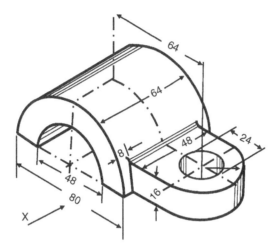

FIGURE 9.25 Assignment P9.9

P9.4 Open problem 5.2 to add datum A on the bottom surface, B on the front view, and datum C at the axis of holes on top view. Add position 0.01 diameter tolerance control frame on the top-hole dimensions, 0.01 flatness control frame on the 2.5 dimension with 0.01 symmetric tolerance. All control frames must be at maximum material condition with A, B, and C datums as references.

P9.5 Open problem 6.1 to add 7 patterned holes, each hole will be positioned at 0.02 positional tolerance from Datum Axis A with 0.01 plus and 0.00 minus tolerances at maximum material condition. Additionally, add 0.02 symmetric tolerance on the center hole, and positional tolerance of 0.01 at maximum material condition from central datum axis A.

CHAPTER 10

Finite Element Analysis (FEA)

10.1 Introduction

Finite element analysis (FEA) is a popular method of solving differential equations in engineering and mathematical modeling. Creo Simulate is an embedded application within Creo Parametric; therefore, there is no need to use a third-party application to conduct simulations. Additionally, Creo Simulate allows to report customized parameters of simulation results. The purpose of FEA in this chapter is to discuss how to study and observe the simulation results that influence design variables such as straight corners vs rounded features or the behaviors of various materials on the same product.

CHAPTER OBJECTIVES

After completing this chapter, you should:

- Understand FEA.
- Install Simulation Package
- Apply Materials to Parts
- Apply Displacement Constraints
- Apply Load Parameters
- Edit Display View Windows
- Interpret Analysis Outcomes

- It is important because we can determine the material, a part will need to be made of for it to accomplish its task. We could find a potentially cheaper material that is strong enough to still produce the part with increasing profits for each part, etc.
- Obviously before you can do an FEA, you have to have a part designed (3D designing will not be covered).
- You will also have to have access to Creo Simulate.

Creo Parametric Modeling with Augmented Reality, First Edition. Ulan Dakeev.
© 2023 John Wiley & Sons, Inc. Published 2023 by John Wiley & Sons, Inc.

Before we get started with the exercise problem, we need to verify that simulate light is installed with our Creo Parametric. If you are using commercial or university-provided Creo Software, it will come with the simulation package installed by default. However, if you have installed the free student version of the software, the package does not install automatically. Therefore, unless you have installed in the beginning, we will need to run the installation package one more time. Because the software is already installed, you do not need to provide new license number to add the package.

10.1.1 Simulation Lite Add-On Installation

- Double-click your downloaded installation file shown in Figure 10.1 (if you deleted it, you can redownload it from ptc using your student account for free).
- On the installation assistant options, select "Upgrade existing software" radio button and click next (Figure 10.2) → Accept the software license agreement on the next screen.

Name	Date modified	Type	Size
install	6/13/2022 12:30 PM	File folder	
pim	6/13/2022 12:31 PM	File folder	
ptcsh0	6/13/2022 12:31 PM	File folder	
export_affirmation	6/13/2022 12:30 PM	Microsoft Edge P...	653 KB
install_license_server	6/13/2022 12:30 PM	Application	173 KB
License_Administration_Guide	6/13/2022 12:30 PM	Microsoft Edge P...	2,479 KB
msgerror	6/13/2022 1:49 PM	Text Document	16 KB
ReadThisFirst	6/13/2022 12:30 PM	Microsoft Edge P...	182 KB
setup	6/13/2022 12:31 PM	Application	173 KB
setup-trial	6/13/2022 12:31 PM	Application	173 KB

FIGURE 10.1 Installing from Creo Package

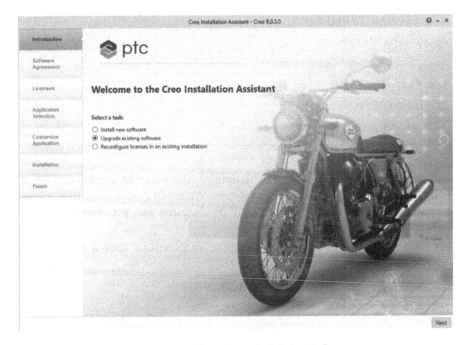

FIGURE 10.2 Select Upgrade Existing Software

- Continue to next screens until you see "Customize Application" (Figure 10.3) →
  Under "Extensions", select "Creo Simulate" (Figure 10.3) → Click Install.
- Next time you start Creo parametric, you will have Creo Simulate Lite under the
  applications.

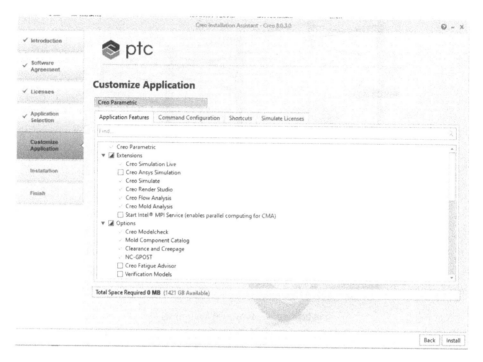

FIGURE 10.3   Check Creo Simulate

E1   **EXERCISE 1** | Workout Rig J-Cup Analysis

In this exercise, we will develop J-Cup part (red color, Figure 10.4) to analyze how much
deformation the part will show when we apply various forces. We will apply two different materials with the same load for the first part of the exercise, and we will make a
slight design change to review its effect on the part.

- Using the dimension values on Figure 10.5, develop J-cup part in metric units.

Simulation 1: Material Comparison

In this exercise, we will apply two different materials to observe the differences between them on the final prototype of the product and draw conclusion about them.
Although we can select number of various outcome metrics, we will be observing three
magnitudes.

(continued)

FIGURE 10.4 Workout Weight Rig

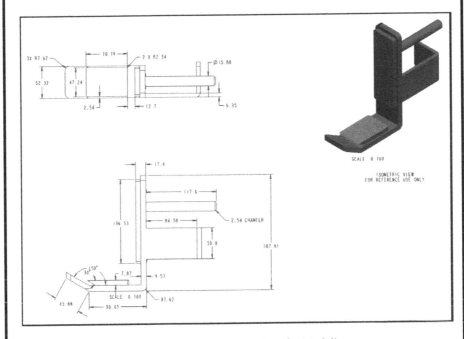

FIGURE 10.5 J-Cup Dimensions for Modeling

| **E1** | **EXERCISE 1** | Workout Rig J-Cup Analysis *(continued)* |

- Start Creo Parametric and open the J-Cup model → Click "Applications" on the Menu (Figure 10.6).

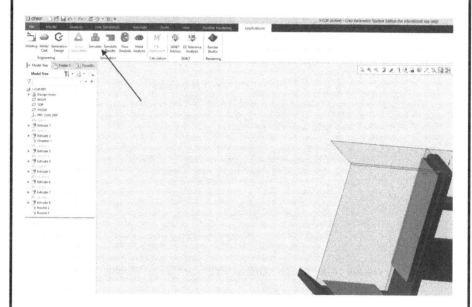

FIGURE 10.6 Simulate Application

You will find various applications related to the analysis depending on the software version installed on your computer. Since we are working with the free student version, we will work with simulate lite, which can be reinstalled (pages 1 and 2).

- Click on "Simulate" (a Process Guide window opens, usually at the bottom right) → Click "Materials" → Click "assign" to assign a new material to the part (Figure 10.7).
- On the Material Assignment window, enter "MySteel" (Figure 10.8a) to name the material you are assigning → Click More (to see loaded materials, Figure 10.8a) → Double-click Legacy-Materials to find Steel (Figure 10.8b) → Double-click "Steel" material to see how it is added to the bottom window (Figure 10.8b).
- Click "Select" to assign the material to the part. Notice that "Material" box (Figure 10.8a) has STEEL material instead of the PTC_SYSTEM → Click OK to finish material assignment.

The J-Cup model now has MySteel tag as an assigned material.

- On the Process Guide window, click "Constraints" → Apply "constraints" underneath to open "Constraints Manager" window (Figure 10.9) → Click "Displacement Manager"

First, we are going to tell Creo simulation what feature of the part is going to be fixed. In this case, since the pin is inserted and holds the majority of load, we will assign it as a fixed feature.

(continued)

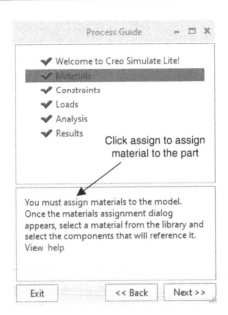

FIGURE 10.7   Assign Material on Process Guide Window

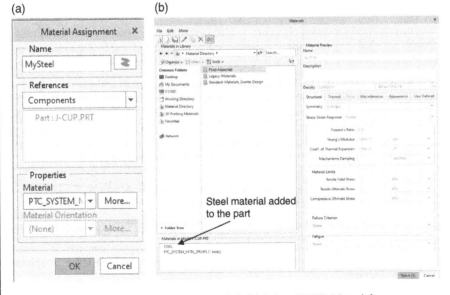

FIGURE 10.8   (a) Assign Material, (b) Select STEEL Material

EXERCISE 1 | Workout Rig J-Cup Analysis *(continued)*

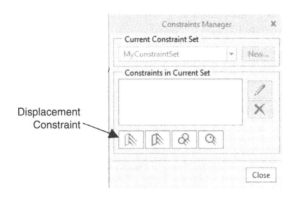

FIGURE 10.9   Constraint Manager Window

- On the "Constraint" window (Figure 10.10) → Enter "FixedPin" or any other name you like for this constraint name → Click on the Pin to indicate what surfaces of the feature will bear the fixed constrain (Figure 10.10) → Apply "Fixed" constraints for both Rotation and Translation.

Note: Because we do not want the pin to have any rotational or translational effects, we will leave them on fixed (Figure 10.10). This will allow us to control the fixed feature. Sometimes when you select a round feature, especially holes, only half of it may show selected. However, the constraint window shows that both are selected.

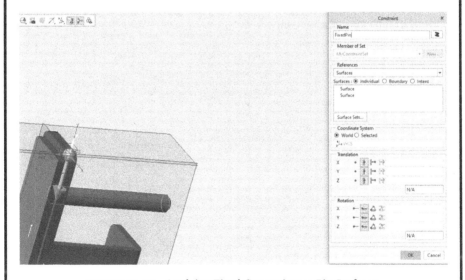

FIGURE 10.10   Applying Fixed Constraints to Pin Surfaces

(continued)

E1 **EXERCISE 1** | Workout Rig J-Cup Analysis (*continued*)

- Click OK to accept the constraint → Visualize how the Fixed constraints are applied on the part (Figure 10.11).

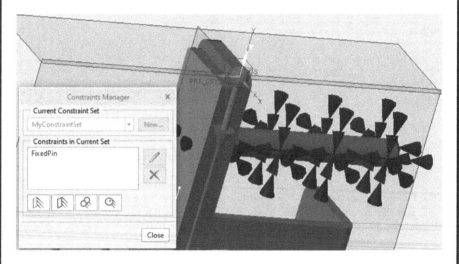

FIGURE 10.11 Visual Representation of Fixed Constraints on the Pin Feature

- Click Loads on the Process Guide window (we will add two loads: 1-a gravitational load 9.8 m/s², and 2-weight load we can rest on the j-cup).
- On the Load Manager (Figure 10.12), click on "gravity load" to add gravitational load (don't forget the units).

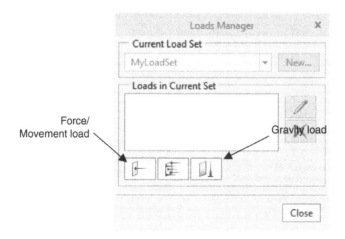

FIGURE 10.12 Loads Manager Window

- Enter gravitational load name (Figure 10.13) ➔ Ensure the gravity's direction is pulled toward the earth (that is negative sign) and units at meter per second square ➔ Click OK when done.

FIGURE 10.13   Gravitational Load Values and Units

- Click on the Force/Movement Load button (Figure 10.12) ➔ Enter the load name ➔ Select the flat surface, where the bar with weights rest (Figure 10.14) ➔ Ensure the direction of the force, enter 500 lbs (for this example) as the load to rest on the j-cup ➔ Click OK when done.
- Notice how the load is visually represented in the work window (Figure 10.15) ➔ Close when done.
- Click "Analysis" on the Process Guide window ➔ Click "Run" to start analysis (It may take a little bit depending on the processor of your computer) ➔ When complete, close the "Run Status" window.
- Click "Results" on the Process Guide window ➔ Click "results" to see presented results (Figure 10.16).

Creo simulate presents three windows by default. We can change these windows to select parameters we are interested in. Notice that one of the windows is animated to show the deformation status. Next, we will organize the windows for the parameters we are interested in:

- Left-click on the third (right window) to activate ➔ Click Edit on the Ribbon (Figure 10.17)
- Under Quantity tab, ensure Stress and psi are selected.
- Click OK and show to display new parameters.

(continued)

E1 **EXERCISE 1** | Workout Rig J-Cup Analysis (*continued*)

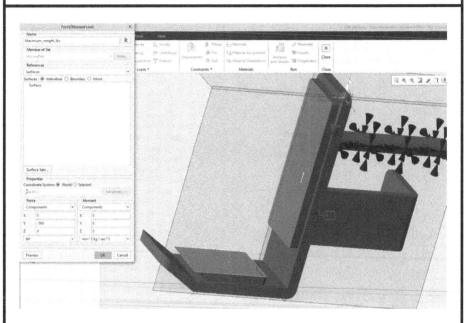

FIGURE 10.14 Entering Weight Load

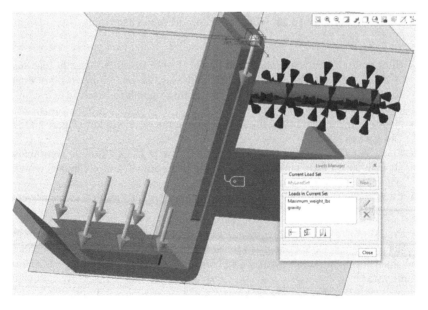

FIGURE 10.15 Visually Inspect All Added Constraints

E1 **EXERCISE 1** | Workout Rig J-Cup Analysis *(continued)*

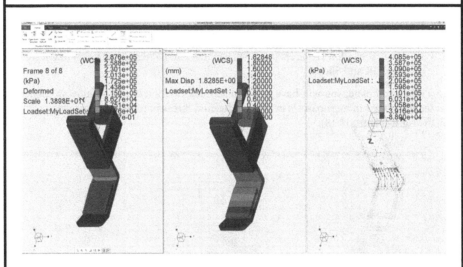

FIGURE 10.16   Analysis Report Windows

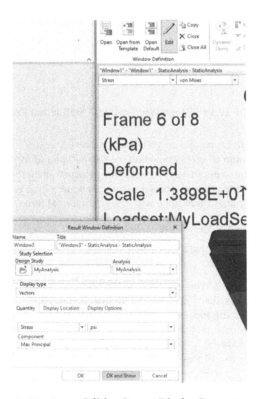

FIGURE 10.17   Editing Report Display Parameters

(continued)

E1 **EXERCISE 1** | Workout Rig J-Cup Analysis *(continued)*

- For the middle window, ensure Stress and Mpa (megapascals) parameters are selected → Click OK and Show.
- For the first window (left), ensure Displacement and inches parameters are selected → Click OK and Show.

When complete, you will have an animated deformation report window in inches on the left, Stress in megapascals in the middle, and Stress in pressure square inch (psi) on the right (Figure 10.18):

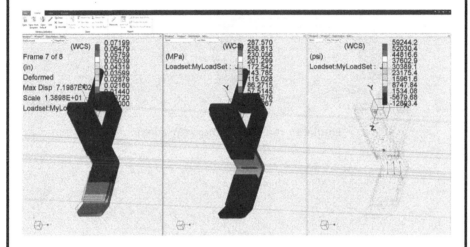

FIGURE 10.18 Three Windows with Simulation Results

You can report these results on your own; however, in summary, we see that the j-cup deforms maximum of 0.7199 in. when we rest 500-pound force (left window). We also see that maximum stress energy exerted to the deformation of the part is 287.570 Mpa (middle window), and maximum stress pressure per square inch is at 59 244.2 psi (right window) for the same amount of force with the gravitational force applied.

- Save the simulation result windows and close.

Because we want to compare how different materials behave under the same applied force, we will change the material constraint and leave the rest of the parameters same.

- On the Process Guide window, click on the Materials constraint → Click "assign" and apply Stainless Steel (SS) from the library (Figure 10.19).
- Click Analysis on the Process Guide (we need to run the analysis one more time, otherwise previously analyzed results will display) → Click Run (wait until Creo completes the analysis).
- Click Results on the Process Guide window → Click Results to display three (by default) simulation result windows for a new material.

  Note: Ensure you edit each window to represent same parameters as for the previous report results (Figure 10.20).

Similarly, visualize the simulated analysis report to see the differences between two materials. We remember that maximum stress in psi for the steel material was 59 244.2 psi, while we observe 60 802.2 psi on the right window. So, which one of the

FIGURE 10.19   Assigning SS Material to the Part

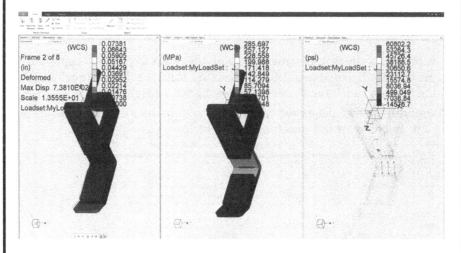

FIGURE 10.20   Simulation Results for Stainless Steel Material

materials do you think is better for manufacturing this part? Surely, the material that can bear more stress, which will be less brittle. This would mean less warranty work and longer life for the product. We can also review the other two windows and generate a detailed report for our organization. Sure, there are other factors that may impact the use of the proposed material such as the cost of stainless steel might be higher than the steel or galvanized steel or carbonized steel etc. If the organization wants to stick with steel, then you can change the design of the part (add rounded feature instead of 90-degree L shape) and see if the analysis reports for the same material makes any difference.

Design Revision for FEA

In the previous exercise, we observed how much deformation occurs when a certain load is applied. Additionally, the analysis report showed where the part will have the most fragile moment (middle window, Figure 10.20). In this exercise, we will add a round feature to the part to see if there is going to be any difference between the two different designs.

- Open the J-cup model unless it is already open → Add 10 mm round feature to the corner (Figure 10.21).

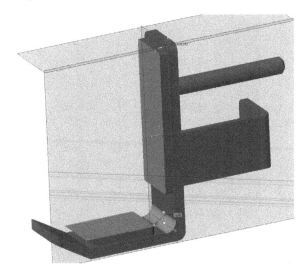

FIGURE 10.21 Design Modification: Add 10 mm Round Feature

- Repeat the analysis with same constraints and loads in previous step. Note that the last material used was stainless steel (SS), leave the material unchanged (do not forget to run the analysis before viewing the results).
- Verify all the parameters and window report units are identical to the previous steps and compare the outcomes (Figure 10.22).

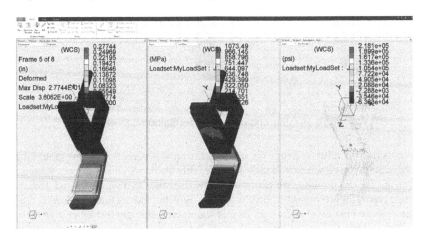

FIGURE 10.22 New Report with the Round Feature

Notice that the deformation displacement has changed from previously 0.07381 to 0.27744 in. when applied same force and constraints. Additionally, we see the stress in both megapascals, and the psi have changed. So, which of the designs do you think is better? Adding more material with round feature to the part or keeping the original design for the same stainless-steel material?

As an independent work, try increasing the diameter of pin to larger dimension and reduce it with the same amount. Run comparative analysis for all three and observe what differences you see. Do not forget to keep same constraints and to report same units and parameters.

10.1.3 Design Revision from FEA

Let us now consider a scenario where the Creo Simulate can guide us through how we can save material on our production. If there is less material to be machined, less time and finances will be needed for the job, which, in turn, saves both time and money for the organization.

- Open the J-cup model unless it is already open ➔ Run the latest simulation and show results ➔ Close two windows on the right to leave the animation window ➔ On the Animated window, change parameters to Stress and Max Principal (Figure 10.23).

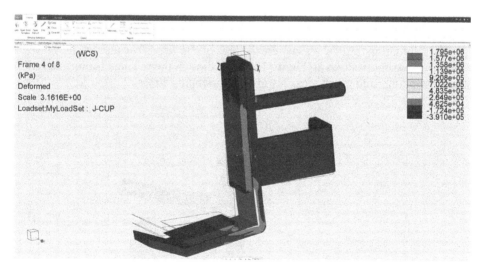

FIGURE 10.23   Keep the Animated Window and Change Parameters

- Click Edit button on the ribbon ➔ Under "Display Options", uncheck "Animate" (Figure 10.24).

FIGURE 10.24 Removing Animation from the Main Display Window

This will remove animation on the main window.

- Click OK and show to display the static stress analysis window.

The simulation result shows where the area of concern is because the part reveals where the maximum stress is (Figure 10.25). Although the red area is our maximum stress, the orange and green areas are also experiencing stress.

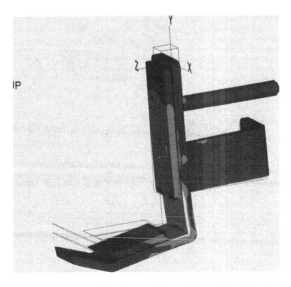

FIGURE 10.25 Maximum Stress Exerted on the Part

- Right-click and hold the mouse button → Select "New Cutting/Capping Surface" (Figure 10.26a) → On the "Results surface" window (Figure 10.26b), change Type to "Capping surface" → Change Define by "Isosurface" → Click "Apply".

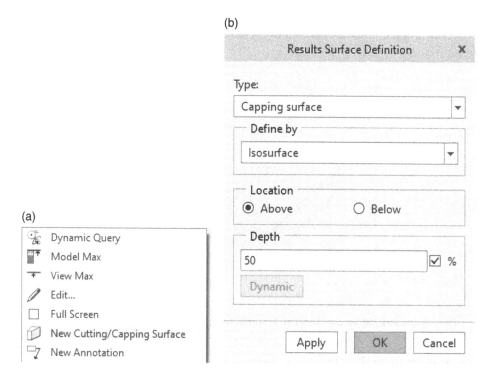

FIGURE 10.26   (a) New Cutting/Capping Surface, (b) Change Type and Define by

Notice that "Dynamic" is inactive, it will be activated only after you click "Apply" button.

- Click "Dynamic" button → Left-click (hold) and move your mouse pointer to see how the part material changes (Figure 10.27).

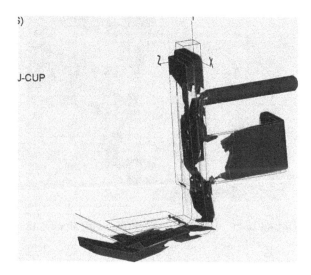

FIGURE 10.27   Move Mouse Pointer to Observe Part Material Dynamically

We observe that the blue areas are the parts, where no stress is exerted. This tells us that we can do some modifications to save material and machining time.

- Close the simulation result window.
- Introduce a cut out at the slanted tip of the part (Figure 10.28). Because we are only analyzing, you can sketch freely to remove material.

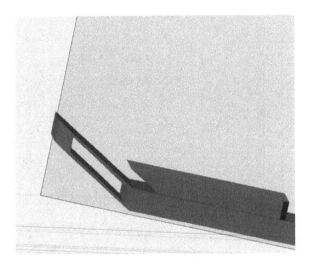

FIGURE 10.28 Remove Material from the Front Slanted Side

- Run the Analysis without any changes. Remember, anytime we make modifications on the model, we need to run the Analysis, so Creo can calculate the updated part and show results.
- Click Results → Close two right windows → Edit displayed parameters for consistency (Stress and Max Principal) → Remove Animation (Figure 10.29).

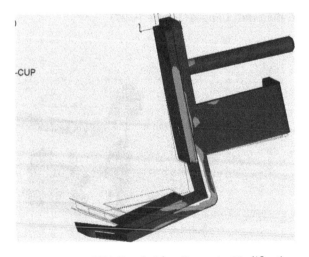

FIGURE 10.29 FEA Result After Geometry Modification

You can observe that the analysis results show that the introduction of a hollow feature for the front slanted part, the overall part is not affected. We can continue to run Capping Surface dynamically to observe if there are other areas we can save on material.

10.1.4 Independent Challenge Activity

As an activity, introduce hollow feature in all blue areas on the J-cup model to see how much material can be saved.

Chapter Problems

P10.1 Model or Open (Chapter 6 problem 1) part with steel material to apply fixed constraints on all screw holes (there are 4 of them), apply gravitational load to the model, apply 75-kg downward load to the center hole. Observe the simulation results and interpret your findings. (A) Investigate how the part will behave if the material is changed to bronze with the same loads and constraints. (B) Where can you introduce design changes to improve the material with stainless steel material (Figure 10.30)?

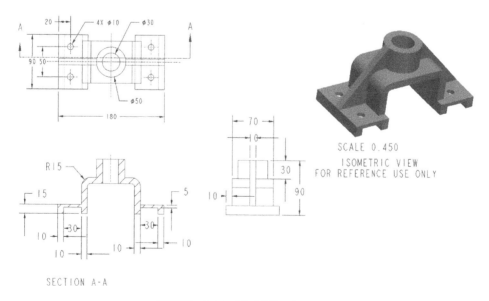

FIGURE 10.30   Model Bracket_0419

P10.2 Model or Open (Chapter 6 Problem 4) part with galvanized steel (Standard materials → Ferrous metals) to apply 200 lb force to the surface from X arrow. Keep the notch feature fixed. Interpret your analysis report, suggest where you can introduce design improvements (Figure 10.31).

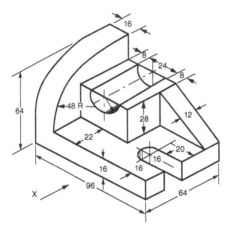

FIGURE 10.31 Rod_Support_1210

CHAPTER 11

Mechanism Assembly

11.1 Introduction

The purpose of an assembly is to establish relationships between parts and to see how various constraints impact their relationship. Because the final product may also be a moving assembly, we can introduce joint connections between the moving parts to observe how the product may behave. You can consider the mechanism assembly and animations as a simulation that provides motion analysis for proper assembly designs and catch problems early on. In Creo mechanism assembly application, there are four basic joint connections: 1-Pin, 2-Slider, 3-Cylinder, and 4-Planar.

Pin joint connection is applied on an axis to provide rotational motion along the axis it is aligned to. Slider connection needs pin connection for an axis, then align or mate a flat surface to the assembly. The slider provides a translational motion on the aligned surface. Cylinder connection allows both rotational and translational motion on the axis aligned to the assembly. Planar connection provides a perpendicular rotational motion to the aligned surface and additional two translational motions (Figure 11.1).

CHAPTER OBJECTIVES

After completing this chapter, you should:

- Understand Mechanism Assembly.
- Use Mechanism Constraints
- Add Servo Motors
- Analyze in Playback
- Save Animated Video File

Creo Parametric Modeling with Augmented Reality, First Edition. Ulan Dakeev.
© 2023 John Wiley & Sons, Inc. Published 2023 by John Wiley & Sons, Inc.

FIGURE 11.1 Nut Cracker

 EXERCISE 1 | Wheel Support Mechanism
(Servo Motor)

Some of the assembly components in the mechanism will be assembled in traditional
way. Therefore, go ahead and model all seven: 1-Base, 2-Support, 3-Wheel, 4-Bushing,
5-Washer, 6-Shoulder Screw, and 7-Nut parts shown on Figure 11.2. For references, the
assembly of the components are also illustrated on Figure 11.3.

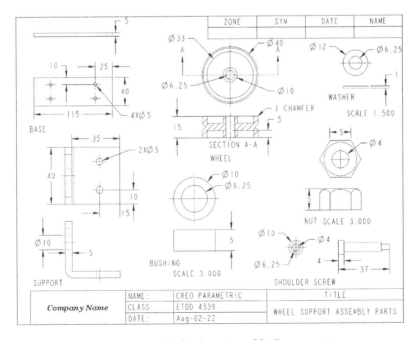

FIGURE 11.2 Mechanism Assembly Components

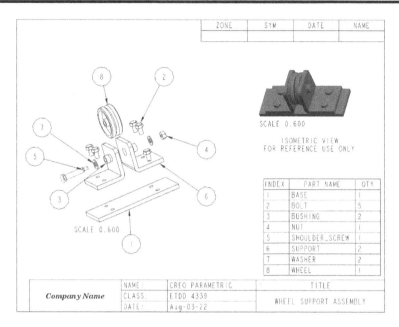

FIGURE 11.3   Wheel Support Assembly Drawing

Challenge: What is the Bolt diameter? (Model the bolt for later assembly).
Once the support wheel parts are modeled, we can start the assembly process:

• Start Creo (unless already running) ➔ New ➔ Assembly ➔ Enter "Washer_
Support_Mechanism" name ➔ Uncheck default (since our parts are in millimeters,
we will develop the assembly in mm, Figure 11.4).

FIGURE 11.4   Starting a New Assembly

- Select mmns_asm_design_abs template for the assembly
- Click Assemble on the menu → Select Base from your working directory → Define "Default" constraint type (Figure 11.5) → OK to accept.

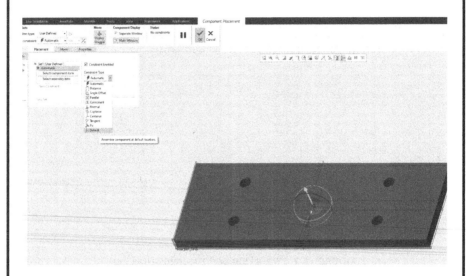

FIGURE 11.5   Default Constraint for Base Part

- Click Assemble tool again → Insert Support part → Assemble as illustrated (Figure 11.6) → OK to accept.

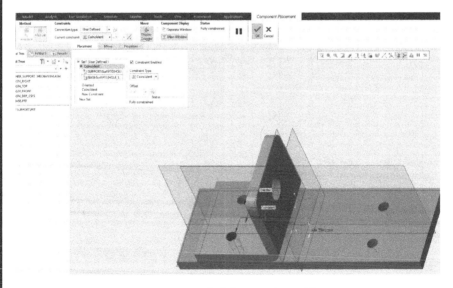

FIGURE 11.6   Left Support Assembly

- Using the same technique above, assemble the right Support (Figure 11.7).

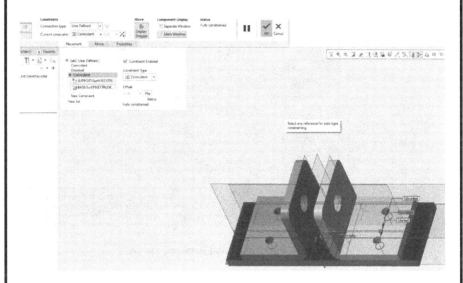

FIGURE 11.7 Right Support Assembly

- Click Assemble ➔ Insert Shoulder Screw part ➔ Click "Connection Type" drop down ➔ Select "Pin" connection (Figure 11.8).

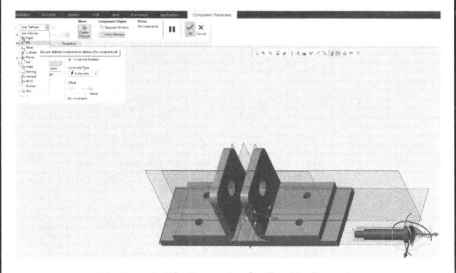

FIGURE 11.8 Pin Connection for Shoulder Screw Part

(continued)

- Select Axis of the Shoulder Screw ➔ Select Axis of the Support part (Figure 11.9).

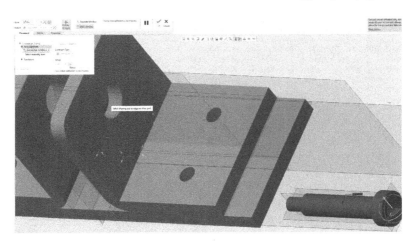

FIGURE 11.9 Pin Connection Between Two Axes

Notice that the axis will lock two degrees of freedom (X and Y). To complete the assembly:

- Select the surface under the Shoulder Screw's cap and the Support part face with coincident constraint (Figure 11.10) ➔ OK to accept.

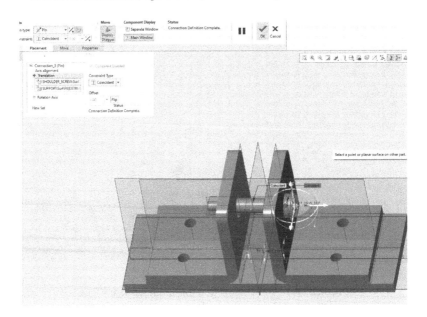

FIGURE 11.10 Shoulder Screw Surface Coincident Constraint

- Select Drag Components tool and try to rotate the Shoulder Screw part. If everything is correct, the part will spin (Figure 11.11) ➔ Close "Drag" menu when done inspecting.

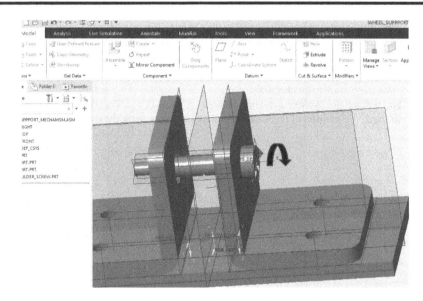

FIGURE 11.11 Rotate Shoulder Screw with Drag Component

- Assemble the Wheel axis to the Shoulder Screw axis and two flat surfaces until the wheel is seated properly (Figure 11.12) → OK when done (You can drag component to observe how wheel is also moving with the Shoulder Screw together).

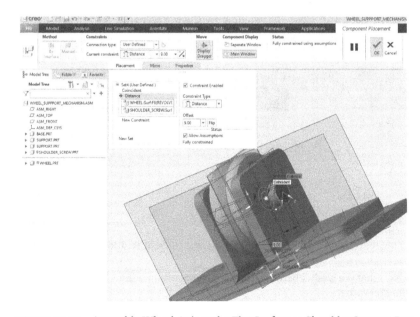

FIGURE 11.12 Assemble Wheel Axis and a Flat Surface to Shoulder Support Part

(continued)

Note: You must constrain the wheel to the pin only. If you constrain wheel axis to the shoulder screw and a flat surface of the wheel to the support part, the assembly will not be complete. Only assembling the wheel to the shoulder screw part will complete the constraints.

• Click Save to save the assembly.
You will see the "Model is not regenerated" (Figure 11.13). This note shows that we changed the position of the wheel and shoulder screw with the drag component.

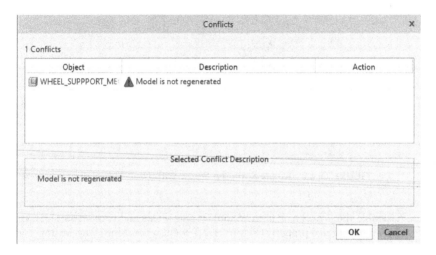

FIGURE 11.13   Model Is Not Regenerated Conflicts Window

• Click Cancel to close the window → Click Regenerate button on the menu (Figure 11.14) → Click Save to save the assembly.

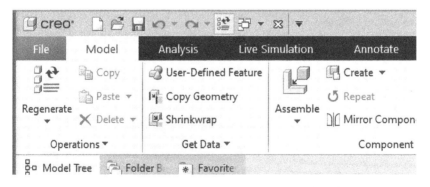

FIGURE 11.14   Regenerate to Accept Current Position of Parts

Note that the washer between the Shoulder Screw and the support part is missing. To add the washer:

• Click Edit Definition on the Shoulder_Screw part on the model tree → Select "Translation" constraint → Change "Constraint Type" from Coincidence to Distance → Enter 1 (washer thickness) in the Offset box (Figure 11.15) → OK when done.

E1 **EXERCISE 1** | Wheel Support Mechanism
(Servo Motor) *(continued)*

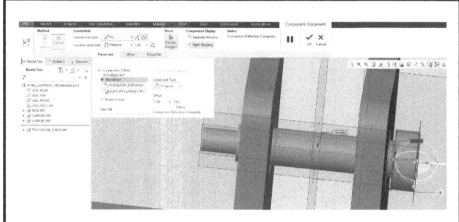

FIGURE 11.15 Editing Assembled Component

- Assemble ➔ Insert Washer part ➔ Axis constraint to Shoulder Screw's axis ➔ Flat surface constrain to Shoulder Screw's flat surface under the cap.
- Assemble the second washer on the other side of the Shoulder Screw (Figure 11.16) ➔ Assemble the nut to complete the components (Figure 11.16).
- Assemble bushings between the support and washer on both sides.
- Insert 4 bolts to fasten the support parts to the base.

FIGURE 11.16 Complete Assembly

11.1.1 Mechanism Animation

For the rotational animation of the Wheel_Support_Assembly practice, we will add a servo motor, which enables to control desired position of the component, its velocity, and acceleration over time. To apply a servo motor, we need to select Applications (we used this tab before for FEA) ➔ Mechanism (Figure 11.17).

- Click Applications ➔ Mechanism (Figure 11.17).

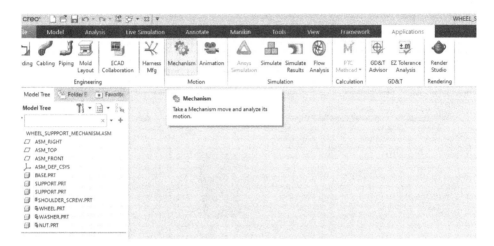

FIGURE 11.17 Mechanism Application

Notice that the ribbon tools have changed, and an orange rotational connection is present on the pin connection we introduced for the Shoulder Screw earlier (Figure 11.18).

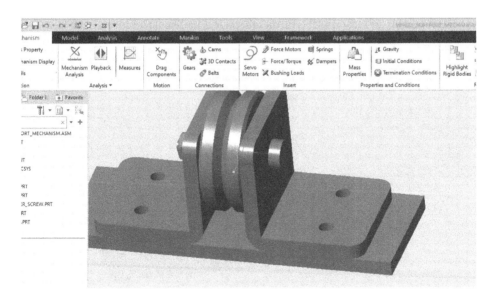

FIGURE 11.18 Servo Motor Connection

- On the Mechanism, click Servo Motors ➔ Select servo connection to place the servo motor on pin connection (Figure 11.19).

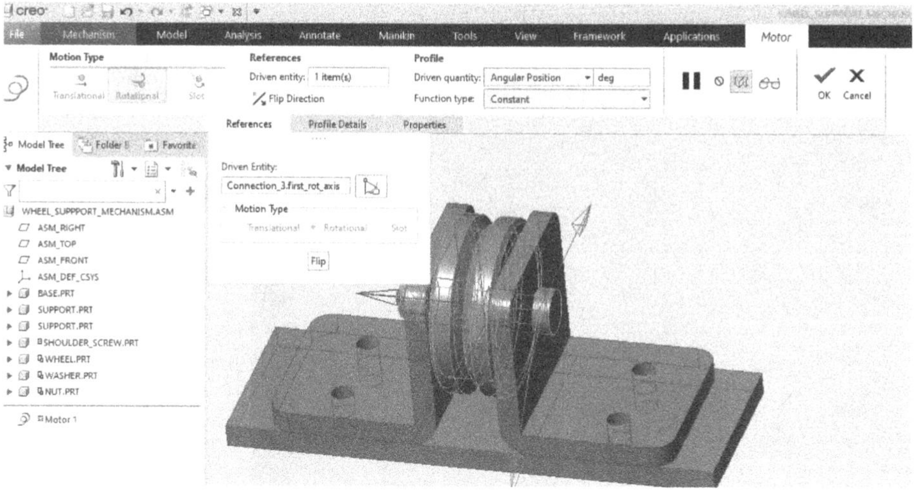

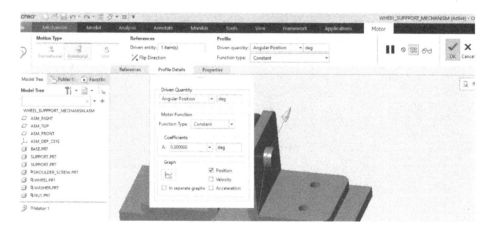

FIGURE 11.19   Servo Motor Placement

- Select the "Profile Details" tab (Figure 11.20).

FIGURE 11.20   Profile Details

- Click "Driven Quantity" drop down → Change to Angular Velocity.
- Leave Motor Function as Constant.
- Enter 36 for Coefficients A. This will allow 36° per second rotation (Figure 11.21).

FIGURE 11.21 Changing Profile Variables

Notice that there is a new Mechanism Tree under the Model Tree on the left.

- Click "Analyses" on the Mechanism Tree (Under Model Tree) ➔ Click "Create New Entity" (Yellow Star) ➔ On the Analysis Definition window, Click Run (Figure 11.22) without any changes.

FIGURE 11.22 Analysis Definition Window

The mechanism will analyze the servo motor placed in the assembly.

- →Click OK to accept.
- Select "Playbacks" on the Mechanism Tree → Click Play.
- On the Playbacks window, click "Play current result set" (Figure 11.23).

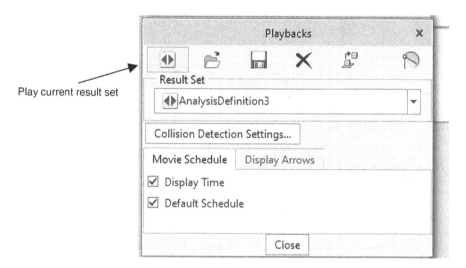

FIGURE 11.23   Playbacks Window

- On the "Animate" window, click Play button to see the animation rotation of the wheel with pin and nut (Figure 11.24).

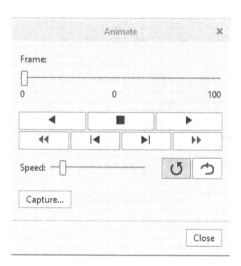

FIGURE 11.24   Animate Window

- Increase the Speed by sliding the Speed slider to the right.
- Click "Capture. . ." to save the animation.
- On the Capture Window (Figure 11.25), select available formats for video output (you can leave mpeg file).

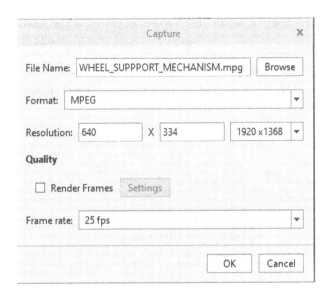

FIGURE 11.25 Animation Video Capture Window

- Click "Browse" to navigate to your working directory and enter a name for the video file.
- Click OK to start saving the animation video.
- Check your working directory to play the animated video → Double click to Play (Figure 11.26).

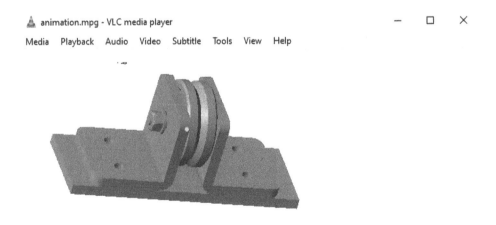

FIGURE 11.26 Saved Animation Video

EXERCISE 2 | Automated Nutcracker Mechanism
(Servo Motor)

For this exercise, we will use some of the Wheel Support Assembly parts and develop new ones to use on the slider mechanism. Unless you have already set up a working directory, move or copy the following parts: Support, Wheel, Bolt (with modifications), Nut, Washer, Bushing into your new working directory and develop the following parts (Figure 11.27): 1-Adjuster_Stop, 2-Bolt (modified), 4-Cylinder, 5-Link, 6-NT_Base, 8-Pin, 9-Piston, 10-Stop_Cap.

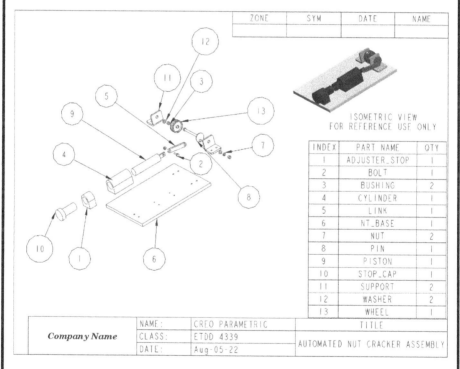

INDEX	PART NAME	QTY
1	ADJUSTER_STOP	1
2	BOLT	1
3	BUSHING	2
4	CYLINDER	1
5	LINK	1
6	NT_BASE	1
7	NUT	2
8	PIN	1
9	PISTON	1
10	STOP_CAP	1
11	SUPPORT	2
12	WASHER	2
13	WHEEL	1

ZONE	SYM	DATE	NAME

ISOMETRIC VIEW
FOR REFERENCE USE ONLY

	NAME:	CREO PARAMETRIC	TITLE
Company Name	CLASS:	ETDD 4339	AUTOMATED NUT CRACKER ASSEMBLY
	DATE:	Aug-05-22	

FIGURE 11.27 Automated Nutcracker Assembly

(*continued*)

- Model the parts from the Figures 11.28 and 11.29

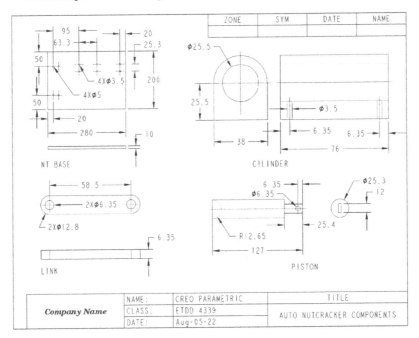

FIGURE 11.28 Automated Nutcracker Assembly Component Dimensions

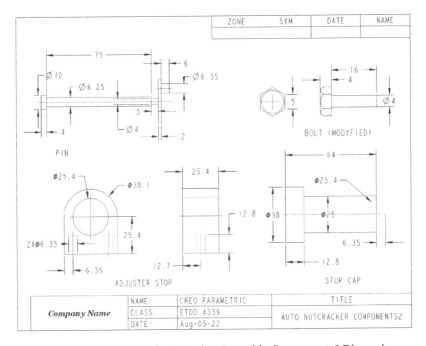

FIGURE 11.29 Automated Nutcracker Assembly Components2 Dimensions

Isometric shaded reference views are illustrated on the fully assembled automated nut cracker assembly on Figure 11.27. The NT_Base and Link (Figure 11.28) parts have their projections below their generic instance, while the Cylinder and Piston parts have their projections on the right. The purpose of this layout is to fit the drawings within the format.

Note that the assembly components are 1 : 1 scale.

The NT_Base and Link (Figure 11.28) parts have their projections below their generic instance, while the Cylinder and Piston parts have their projections on the right. The purpose of this layout is to fit the drawings within the format.

- Start a new assembly in mmns_asm_design_abs units, call it Auto_Nut_Cracker.
- Assemble → NT_Base → Default.
- Using standard assembly techniques, assemble both Supports → Assemble → Insert Pin Part → Using "Pin" connection, constan Pin part's axis to one of the Support's axis → Using Translate constrain, mate two flat surfaces of the Pin part and the same Support part with distance (if you select a different part, the Pin part will stay purple and the assembly will not complete, it is important to use the same mating parts to fully define all necessary constraints).
- Using standard assembly techniques, assemble Wheel part to the middle of the Pin between two Support parts.
- Using standard assembly techniques, assemble both washers and bushings on both sides of the two Support parts → Assemble nut to complete this step (Figure 11.30).

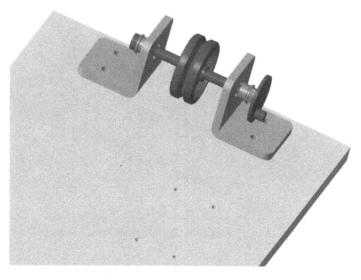

FIGURE 11.30 Assemble Base, Wheel, Supports, Bushings, Washers, and Nut

- Assemble the Link part with Pin connection to the end of the Pin part (Figure 11.31).

(continued)

FIGURE 11.31   Assemble Link Part Through Pin Connection to Pin Part

- Using standard assembly techniques, assemble Cylinder part onto the base.
- Similarly, assemble Adjuster_Stop part onto the base (Figure 11.32).

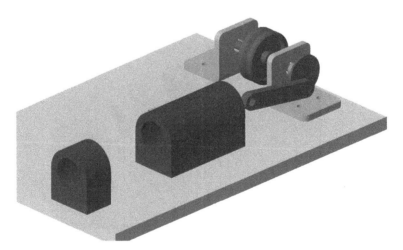

FIGURE 11.32   Cylinder and Adjuster_Stop Parts Assembled

- Assemble (Insert) Piston part → Connection Type "Pin" → Connect Piston Axis to Link's axis → Mate flat surface of the Piston part to Link (Figure 11.33).

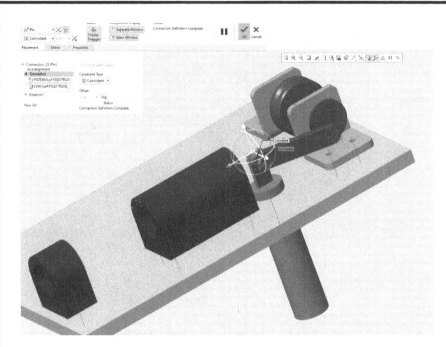

FIGURE 11.33 Pin Connection for Piston and Link

The purpose of the pin connection was to let the piston to rotate as the link rotates. Next, we will introduce new set of constraints, introduce "Cylinder" connection (so the Piston part can slide):

- Click "New Set" on the "Placement" tab to introduce new set of constraints (Figure 11.34).

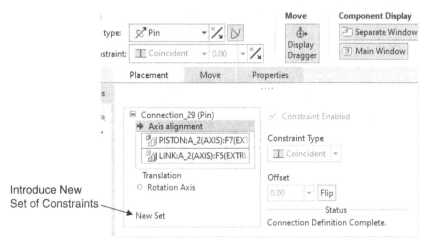

FIGURE 11.34 Introduce New Set of Constraints

(*continued*)

- Connection Type → Cylinder (Figure 11.35).

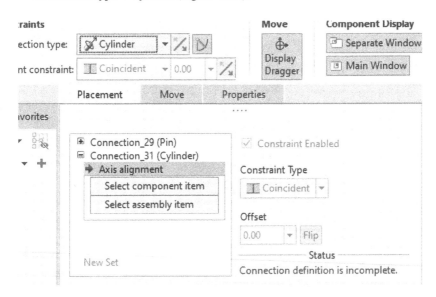

FIGURE 11.35 Select Cylinder Connection Type

- Select Piston's axis → Connect it to Cylinder part's axis (Figure 11.36, if the piston is disoriented, click "Flip") → OK to accept.

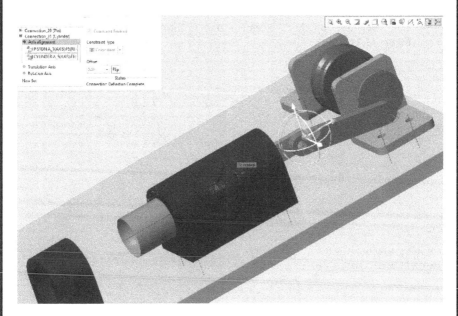

FIGURE 11.36 Piston Axis to Cylinder Axis, Click Flip to Orient Piston If Needed

- Click "Drag Components" tool (hand) → Spin the Wheel part to observe how the piston is sliding within the Cylinder part.
- Using standard assembly techniques, assemble Stop_Cap, Bolts, Nut parts (Figure 11.37).

FIGURE 11.37   Complete Mechanism Assembly with Pin and Cylinder Connections

- Click Applications → Mechanism → Insert Servo Motor to the Pin part (Figure 11.38, to let the Wheel part to rotate).

Insert Servo Motor here

FIGURE 11.38   Servo Motor on Pin Part

- Enter the same Servo Motor parameters as the "Wheel_Support_Mechanism" → Play to observe the animation (Figure 11.39) → Save the video output.

(*continued*)

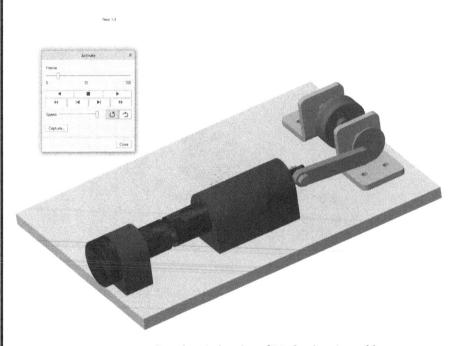

FIGURE 11.39 Complete Animation of Mechanism Assembly

- Save and Close the project.

Chapter Problems

P11.1 Develop a mechanism assembly and insert a servo motor that turns the Swinging_Link assembly's link part from one side of the base to the other to stop before colliding with the base. Capture the animation video of the outcome.

P11.2 Develop pulley support mechanism below and add servo motor to spin the pulley (Figure 11.40).

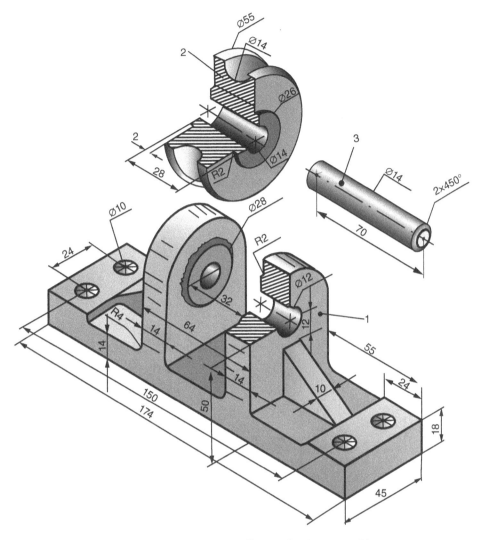

FIGURE 11.40 Develop Pulley Mechanism Assembly

P11.3 Develop vice assembly mechanism below and add servo motor to spin the pulley and slider connection to slide Jaw Face (Figure 11.41).

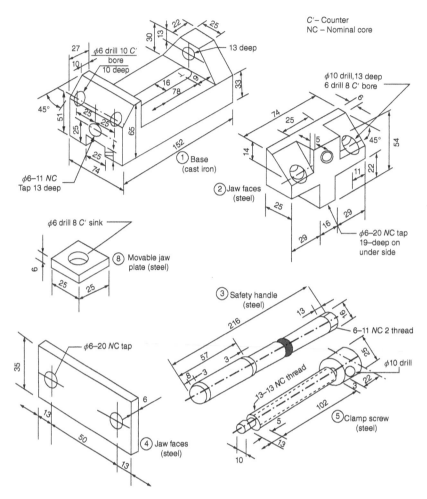

FIGURE 11.41 Develop Vice Mechanism Assembly

CHAPTER 12

Sheetmetal Modeling

FIGURE 12.1 Heavy Duty Joist Hanger

12.1 Introduction

Majority of the products you have encountered are manufactured using sheet metals (Figure 12.1). The sheet metals are used in air-conditioning ducts and furnace flues, air cowls, awnings, canopies, electronic enclosures, roofs, gutters, etc. Every time you receive or ship a product in mailboxes, you are using the sheet metal product. Sheet metal products are punched, stamped, or burned out of flat material and bent to the required shape.

In this chapter, we will learn sheet metal tools, their similarities, and how they differ from each other to product the needed product model. Additionally, we will learn how to generate a sheet metal drawing on the template we custom developed on Chapter 5.

Creo Parametric Modeling with Augmented Reality, First Edition. Ulan Dakeev.
© 2023 John Wiley & Sons, Inc. Published 2023 by John Wiley & Sons, Inc.

CHAPTER OBJECTIVES

After completing this chapter, you should:

• Understand Modeling Sheet Metals.

• Use Bend Components

• Use Flat Components

• Apply Variable Options

• Develop drawings for Sheetmetal parts

E1 **EXERCISE 1** | Bolt on Bracket

For this exercise, we will develop a heavy-duty metal bracket out of a 0.25-in. plate (Figure 12.2). Notice that the drawing illustrates traditional orthogonal views of the part from three sides. Later, we will also develop a drawing to illustrate the unbent part directly cut from the Sheetmetal. Because this drawing is a communication between various parties, the bent (complete) part drawing may serve as a clear communication too.

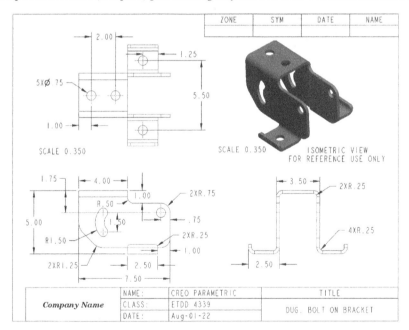

FIGURE 12.2 Bolt on Bracket Drawing

• Start Creo Parametric (unless it is already running) ➔ New ➔ Sheetmetal (Figure 12.3) ➔ OK.

Note: Creo Parametric is defaulted to imperial units; hence, we left the "Use default template" unchecked. If you have changed the defaults to metric units, go ahead and start the part in inches.

• Inspect the Interface Figure 12.3). Notice that the ribbon has slightly different tools and only three tools are active: Extrude, Planar, and Boundary Bend.

E1   **EXERCISE 1** | Bolt on Bracket (*continued*)

FIGURE 12.3   Start Sheetmetal Part in Imperial Units

Extrude in Sheetmetal (Tab)

The extrude function in Sheetmetal tab (Figure 12.4) is used to draw open sketches with lines, which will turn into a Sheetmetal with the designated thickness. Let us quickly practice extrude in Sheetmetal tab:

- Click "Extrude" button ➔ Select Plane to sketch ➔ Sketch 5×5 line (Figure 12.5a) ➔ Click OK ➔ Enter 0.05 thickness and 5 depths for the Sheetmetal (Figure 12.5b).
- Click OK or Cancel (this is the end of Extrude in Sheetmetal Practice).

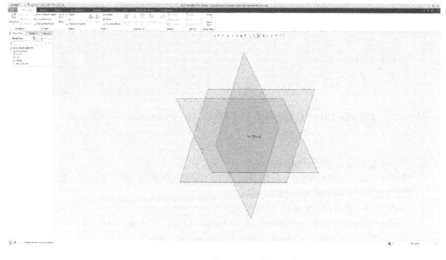

FIGURE 12.4   Sheetmetal Interface

(*continued*)

(a)

(b)

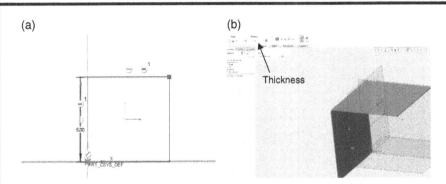

FIGURE 12.5 (a) Sketch Line Trajectory, (b) Enter Geometry Thickness

Planar

This might be little confusing, but the Planar behaves similar to "Extrude" in the model, which lets you sketch a closed geometry and accept the final sketch. However, once the plane is complete, we need extra step to bend the geometry. Let us build the same 5×5×5 Sheetmetal part in the previous practice.

- Click "Planar" ➔ Select a plane ➔ Sketch a 5×5 Rectangle (Figure 12.6a) ➔ OK to accept ➔ Enter 0.05 Sheetmetal thickness (Figure 12.6b).

(a)

(b)

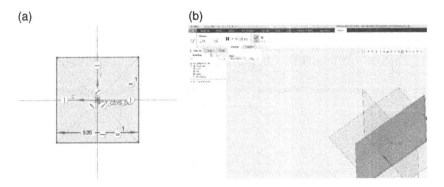

FIGURE 12.6 (a) Sketch 5×5 Plane, (b) Sheetmetal Thickness

- Click OK to accept the plane.
  Notice additional tools are activated once the plane geometry is ready. We will use Flat tool to bend the 5×5 plate.
- Click "Flat" tool (Figure 12.7) ➔ Select one of the edges to start bending (this edge will be the outer edge or radius) ➔ Enter the length of the newly introduced plate.
- Click OK or Cancel (this is the end of Planar Practice).

E1   **EXERCISE 1** | Bolt on Bracket *(continued)*

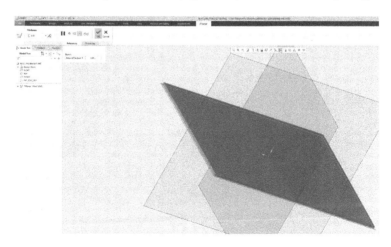

FIGURE 12.7   Enter 5-in. Length

Continue with Exercise 1

Now that you understand the difference between the two tools, there are strengths for each tool use. We will use the combination of both in this exercise. The part is made of 0.25-in.-thick aluminum.

- Click Planar ➔ Select any datum plane ➔ Sketch a 4×3.5 rectangle (top of the part) ➔ OK ➔ Enter 0.25 as plate thickness (Figure 12.8) ➔ Click OK to accept the geometry.

FIGURE 12.8   4×3.5 Plate with 0.25-in. Thickness

(continued)

E1 **EXERCISE 1** | Bolt on Bracket (*continued*)

- Click "Flat" tool and select the edge of 4-in. side → Enter 5-in. length (Figure 12.9).

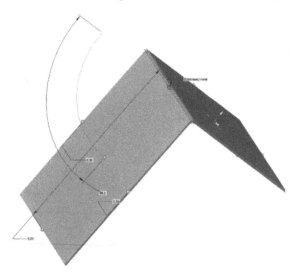

FIGURE 12.9 5-in. Length

Notice that this face is not only 5 in. deep, but also extends to 7.5 in. to the side (Figure 12.2). Therefore, we need to sketch additional feature within the flat.

- Select "Shape" tab within the Flat → Click "Sketch" (Figure 12.10).

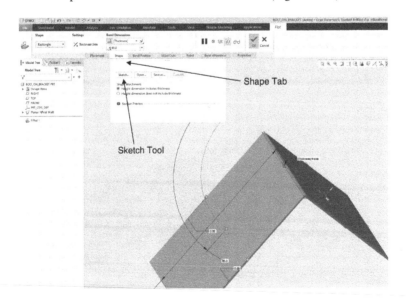

FIGURE 12.10 Sketching New Feature Within Flat Tool

- Sketch additional feature and define strong dimensions (Figure 12.11).

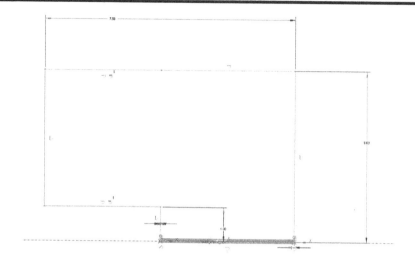

FIGURE 12.11   Sketch Needed Geometry

- Click OK to accept the sketch and inspect the model (Figure 12.12) ➔ Click OK when done.

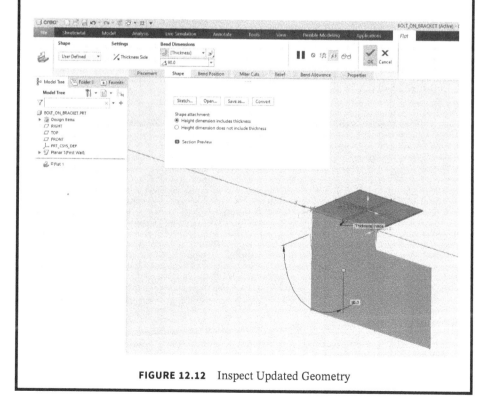

FIGURE 12.12   Inspect Updated Geometry

(continued)

E1 **EXERCISE 1** | Bolt on Bracket (*continued*)

We can continue using Flat tool for additional flange feature directly on the face; however, let us use extrude tool to put it into the practice one more time. Remember, we need to sketch trajectory lines perpendicularly to the newly introduced face (Figure 12.13).

- Click "Extrude" tool ➔ Select the perpendicular side of the face (Figure 12.13) ➔ Sketch 2.5 long line and ➔ 0.75 tall line ➔ Click OK.

FIGURE 12.13 Sketch Trajectory Lines Perpendicularly

If the Sheetmetal is extruding the wrong direction, click on the arrow to flip.

- Enter 2.5 depth for the flange (Figure 12.14).

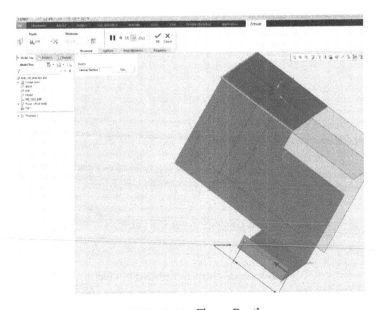

FIGURE 12.14 Flange Depth

E1   **EXERCISE 1** | Bolt on Bracket *(continued)*

Because we started sketching on the right surface, the 2.5 depth starts immediately. However, the specification requires us to have an 1-in. relief from the side (Figure 12.2). Therefore:

- Click Options tab → For Side 2, select Variable → Enter 1 as a value to add relief from the edge, because this cuts from the 2.5 depth, we need to enter 3.5 for Side 1 Variable (Figure 12.15) → Click OK to accept the geometry.

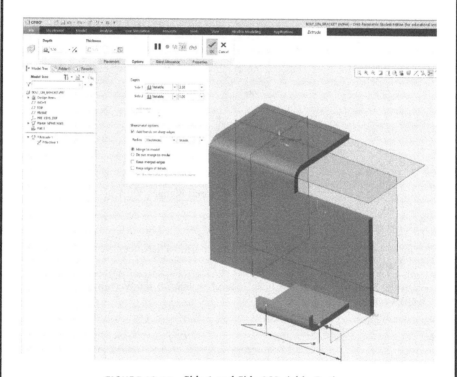

FIGURE 12.15   Side 1 and Side 2 Variable Options

- Select both Flat 1 and Extrude 1 from the Model tree (Hold Ctrl button for multiple selection) → Right Click for Mirror → Select the middle plane to mirror flanges to the opposite side (Figure 12.16).
- Click Model menu tab → Click Extrude → Select the Top View → Sketch four 0.75 diameter holes.
- Select the Side View → Sketch another 0.75 diameter hole → Extrude all the way to the opposite side.
- Similarly, on the same surface, sketch the arced cut outs.
- Add 1.5 Chamfer.
- Add 0.25, 0.5, and 1.25 rounds all around.

(continued)

E1 **EXERCISE 1** | Bolt on Bracket *(continued)*

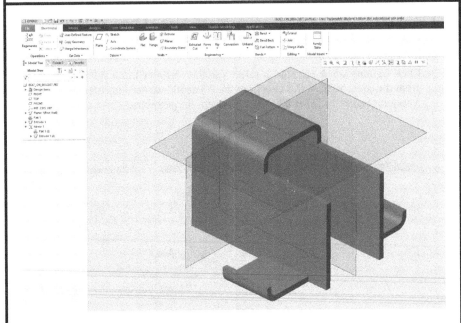

FIGURE 12.16 Mirrored Flanges

- Once complete, save your work (Figure 12.17a).

Since Sheetmetal products are manufactured from a flat sheet, we need to show the unbent shape of the part:

- On the Sheetmetal Tab → Click Unbend → Inspect for Warnings and Errors → Click OK to accept (Figure 12.17b).

(a)   (b)

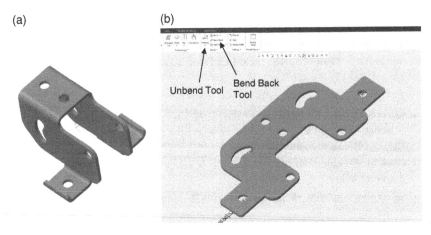

FIGURE 12.17 (a) Complete Bracket Part, (b) Unbent Sheetmetal Part

If you need to bend the part features back, click on the Band Back tool.

E2 **EXERCISE 2** | Corner Bracket

In this exercise, we will cover a few more tools to develop Sheetmetal parts.

The Corner Bracket (Figure 12.18) is made of a gauge 16 brass with 0.050-in. thick Sheetmetal.

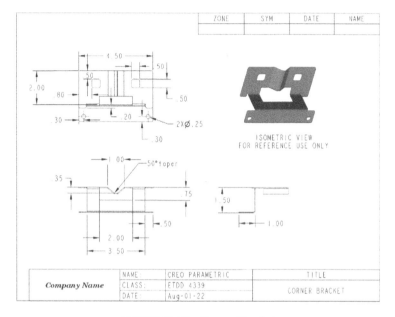

FIGURE 12.18 Corner Bracket

- Start Creo Parametric (unless it is already running) ➔ New ➔ Sheetmetal ➔ Enter "Corner_Bracket" name ➔ Choose imperial units.
- Start the project with Planar tool ➔ Sketch the side face (Figure 12.19).

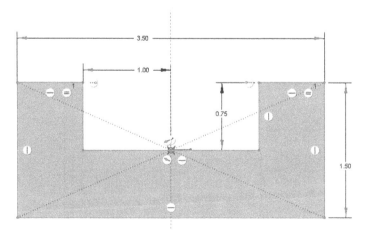

FIGURE 12.19 Side Face Sketch on Planar

(*continued*)

- Ensure the Sheetmetal thickness is 0.05 → OK to accept.
- Introduce a Flat feature (with 0.5 in. from each side) → Click OK to accept (Figure 12.20).

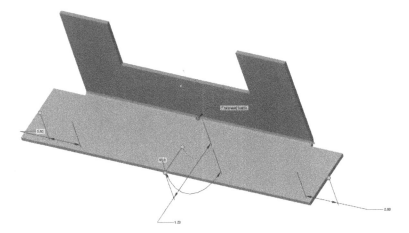

FIGURE 12.20 Add Flat Bend with 0.5 Extensions on Both Sides

Join Command

We can add Flat to the face without any problem. However, the top section of the part on Figure 12.20 is split in between. Therefore, using Flat will not work. Instead, we will introduce a new Planar Sheetmetal (distinct piece) on top of the surface and join the two pieces.

- Select Planar → Sketch the top piece (Figure 12.21) → Click OK to accept.

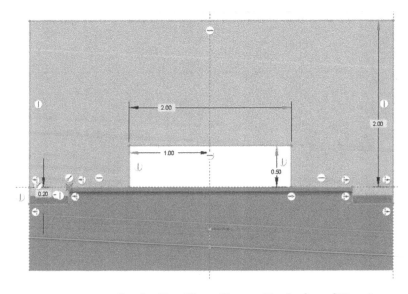

FIGURE 12.21 Sketch a New Planar Piece on Top Surface of Piece 1

Once complete, you should see two distinct pieces (Figure 12.22) at the bottom of the screen. These two distinct pieces are independent from each other at this point. Joining the two pieces will merge into one with all available Sheetmetal tools.

FIGURE 12.22   Independent Distinct Pieces Before Join

- Click "Join" tool → Select Distinct Piece1 → Hold Ctrl Key and select Distinct Piece 2 (Figure 12.23).

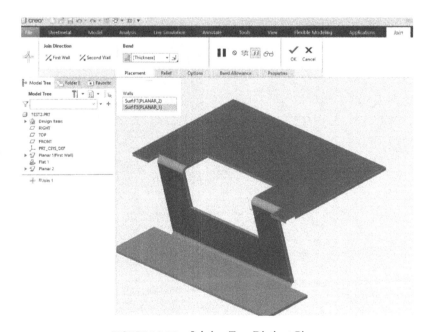

FIGURE 12.23   Joining Two Distinct Pieces

(*continued*)

Before we accept the joint piece,

• Click "Options" → Select "Up to intersection line" radio button selection (Figure 12.24b)

(a)   (b)

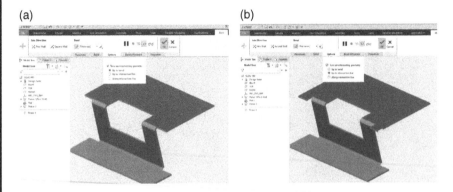

FIGURE 12.24 (a) Join Cuts off the Sides, (b) The Sides Are Back

• Add "Obround" relief to the bend (Figure 12.25) → Click OK to accept.

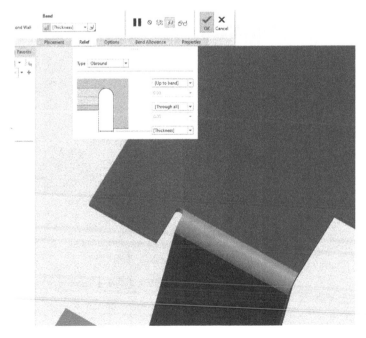

FIGURE 12.25 Obround Relief Added

E2   EXERCISE 2 | Corner Bracket *(continued)*

Form Command

So far, the bends have been working well with the same size Sheetmetal features. However, the Corner_Bracket part has a form feature at the middle (Figure 12.26). If it was a bend feature, the top material would become longer than the joined bottom piece, which is against physics. Trying to bend it will generate error in creo. Hence, we will introduce the Form tool.

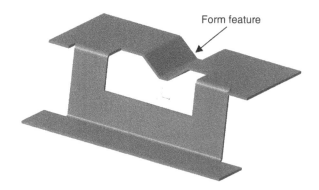

Form feature

FIGURE 12.26   Form Feature

• Click Form drop-down arrow ➜ Select Sketched Form (Figure 12.27).

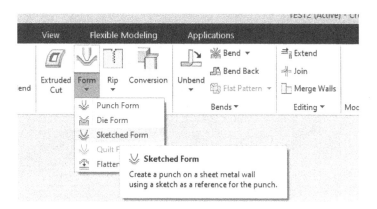

FIGURE 12.27   Sketched Form Command

• Select the Top surface, where we are interested in adding form ➜ Sketch a rectangular shape for Form's width (Figure 12.28) ➜ Click OK to accept.

(continued)

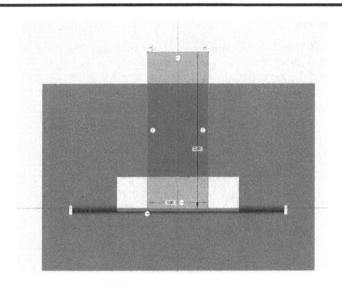

FIGURE 12.28 Sketching Form Width and Height

- Notice that the predefined form is presented.
- Change the form arrow direction if necessary so it is forming downward.
- Click the "Options" tab → Check "Add taper" → Enter the taper value (50° for this part)→Check both "Round sharp edges" (Figure 12.29).

FIGURE 12.29 Tapered Form Feature

- Click OK to accept the form

Add all holes and rounds to complete the Corner_Bracket part (Figure 12.30).

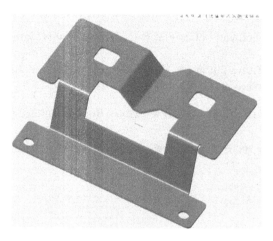

FIGURE 12.30   Complete Part

- Optional, unbend the part to inspect the form feature. Note that the Form did not flatten.
- Bend the part back → Save and close.

12.1.1   Sheetmetal Drawing

Both exercises contain regular complete part isometric view and their orthogonal views with dimensions. Although this is a good practice to represent the final product, you may want to add the flat pattern (unbent view) of the part as well. Since we already have developed the regular drawing (master rep) of the part (Figure 12.31), we will continue to add the flat pattern on the same drawing.

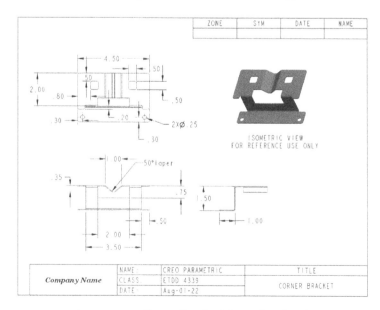

FIGURE 12.31   Corner Bracket Master Rep drawing

If we unbend the part in the model and switch to the drawing view, all views will be replaced with the flat pattern, which is not the purpose of this practice. Therefore, we need to introduce a new representation (rep) of the part and add it to the drawing.

- On the Sheetmetal view, click Flat Pattern drop down (Figure 12.32).

Initially, the features under the Flat Pattern are greyed out and not accessible. To access them, we need to introduce Flat Pattern first.

- Click Flat Pattern (Figure 12.32) ➔ Inspect the Flat Pattern ➔ OK to accept

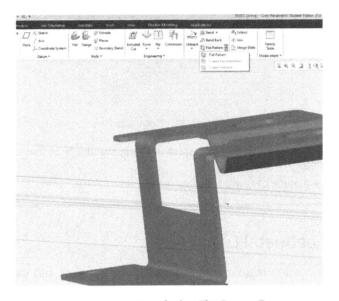

FIGURE 12.32 Introducing Flat Pattern Rep

The Flat Pattern generated a new representation of the model and unbent the part (Figure 12.33).

FIGURE 12.33 Flat Pattern (Unbent)

Next, we need to generate Flat Pattern Rep to save as a standalone representation.

- On the Sheetmetal tab, click Flat Pattern drop down → Select "Create Representation" (Figure 12.34a).
- Click "Create" on the Make Flat Representation window (Figure 12.34b).

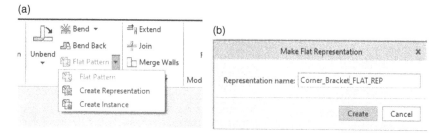

FIGURE 12.34   (a) Flat Pattern Rep Generation, (b) Make Flat Rep Window

- Activate your Master Rep view on the model Tree (move the bottom line one step up to activate the master rep, Figure 12.35).

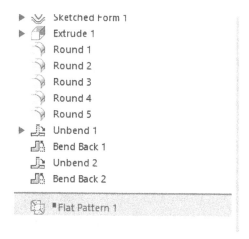

FIGURE 12.35   Activate Master Rep on Model Tree

- Click the View Manager → Simp Rep Tab → Notice you have now the Corner_Bracket_Flat_Rep in the list.
- Switch back to the drawing.
- Under Layout tab, click "Drawing Models".

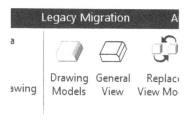

- On the Menu Manager (Figure 12.36) → Select "Set/Add Rep" → Select "Corner_Back_Flat_Rep" at the bottom of the menu → Click Done/Return when done.

FIGURE 12.36 Replace Representation

- Click "General View" → Click on an empty space to add the flat representation of the Corner_Bracket part (Figure 12.37). Notice the part will come with two dimensions.

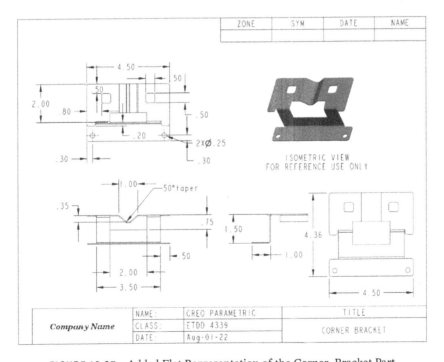

FIGURE 12.37 Added Flat Representation of the Corner_Bracket Part

Chapter Problems

P12.1 Model a Connector Bracket 8222 with 0.125-in. Sheetmetal, where L = 5.5, W1 = 3, W2 = 4, and H = 4 in. Develop a drawing with three orthogonal views, isometric view, and a flat rep of the part (Figure 12.38).

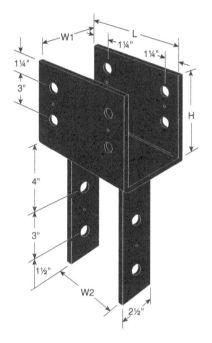

FIGURE 12.38 Connector Bracket 8222

P12.2 Model a Corner Bracket with 0.25-in. plate, where L = 3, W1 = 5, W2 = 5, D1 = 0.5, D2 = 2, D3 = 0.5 and D4 = 2 in. Develop a drawing with three orthogonal views, isometric view, and a flat rep of the part (Figure 12.39).

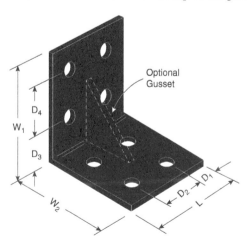

FIGURE 12.39 Corner Bracket

P12.3 Model a Strong-Tie with 0.05-in. Brass Sheetmetal. Add 45-degree chamfer on the top piece and 60-degree chamfer on the bottom piece. Develop a drawing with three orthogonal views, isometric view, and a flat rep of the part (Figure 12.40).

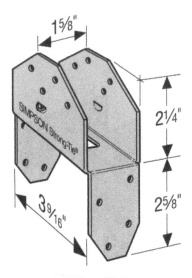

FIGURE 12.40 Strong-Tie

P12.4 Model a Corner Bracket FBGB150RH with 3 mm Stainless Steel Sheetmetal. Add 45-degree chamfer on the top piece and 50-degree chamfer on the bottom piece. Develop a drawing with three orthogonal views, isometric view, and a flat rep of the part; 12 mm holes on the largest face are located 30 mm from side and 20 from bottom. Distance between the holes is 80 mm. The top holes are 12 mm in diameter, 50 mm between centers, and located at the center of the top surface and 25 mm from the straight edge (Figure 12.41).

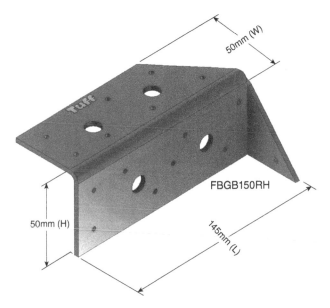

FIGURE 12.41 Corner Bracket FBGB150RH

P12.5 Model a Face-Mount Girder Hanger with 0.129-in. Aluminum with W = 9″, B = 5 1/4″, and H = 20″. Develop a drawing with three orthogonal views, isometric view, and a flat rep of the part. Equally spread M8 bolt holes (Figure 12.42).

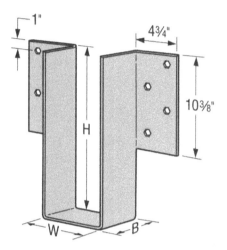

FIGURE 12.42   Face-Mount Girder Hanger

Appendix - Augmented Reality Companion Application and Image Targets

Chapter 2

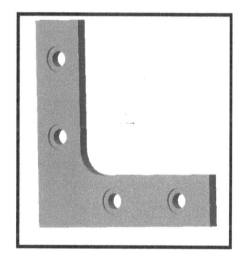

Chapter 3

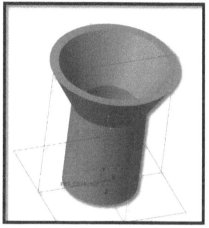

Chapter 4

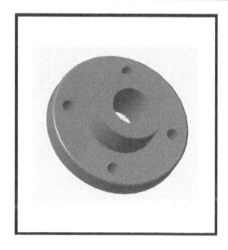

Chapter 5

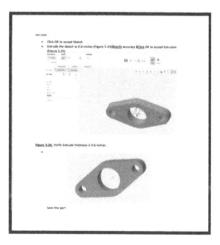

Creo Parametric Modeling with Augmented Reality, First Edition. Ulan Dakeev.
© 2023 John Wiley & Sons, Inc. Published 2023 by John Wiley & Sons, Inc.

Chapter 6

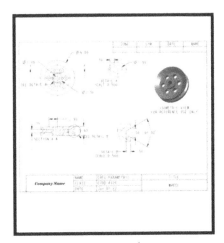

Chapter 7

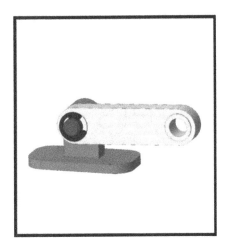

Chapter 9

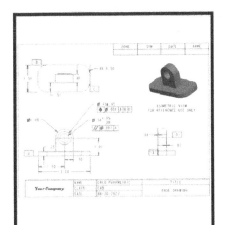

Chapter 10

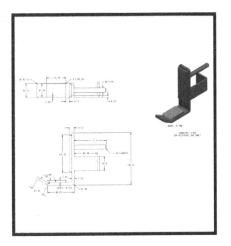

Chapter 11

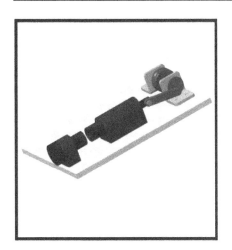
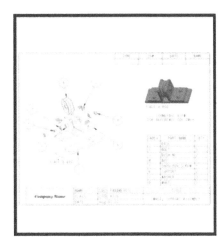

Chapter 12 Bent

Chapter 12 Flat

Index

Creo Parametric Modeling with Augmented Reality, First Edition. Ulan Dakeev.
© 2023 John Wiley & Sons, Inc. Published 2023 by John Wiley & Sons, Inc.